Photograph taken by a prisoner in Ruhleben camp, near Berlin, undated (detail). Maurice L. Ettinghausen Collection of Ruhleben Civilian Internment Camp Papers, Harvard Law School Library, Historical and Special Collections.

Knowing by Ear

....

Listening to Voice Recordings with African Prisoners of War in German Camps (1915–1918)

ANETTE HOFFMANN

DUKE UNIVERSITY PRESS DURHAM AND LONDON 2024

© 2024 DUKE UNIVERSITY PRESS
All rights reserved
Printed in the United States of America on
acid-free paper ∞
Project Editor: Livia Tenzer
Designed by Matthew Tauch
Typeset in MeropeBasic and Comma Base by Westchester
Publishing Services

Library of Congress Cataloging-in-Publication Data
Names: Hoffmann, Anette, [date] author.
Title: Knowing by ear : listening to voice recordings with African prisoners of war in German camps (1915-1918) / Anette Hoffmann.
Other titles: Sign, storage, transmission.
Description: Durham : Duke University Press, 2024. | Series: Sign, storage, transmission | Includes bibliographical references and index.
Identifiers: LCCN 2023028134 (print)
LCCN 2023028135 (ebook)
ISBN 9781478030027 (paperback)
ISBN 9781478024842 (hardcover)
ISBN 9781478059028 (ebook)
Subjects: LCSH: Prisoners of war—Germany—History—20th century—Archival resources. | Africans—Germany—History—20th century—Archival resources. | World War, 1914–1918—Prisoners and prisons, German | Prisoners' songs—Germany. | Sound recordings in ethnology—Germany. | BISAC: HISTORY / Africa / General
Classification: LCC D627.G3 H565 2024 (print) | LCC D627.G3 (ebook) | DDC 940.54/7243—dc23/eng/20230914
LC record available at https://lccn.loc.gov/2023028134
LC ebook record available at https://lccn.loc.gov/2023028135

Cover art: Photograph taken by a prisoner at Ruhleben camp, Germany, undated (detail). Maurice L. Ettinghausen Collection of Ruhleben Civilian Internment Camp Papers, Harvard Law School Library, Historical and Special Collections.

Contents

Note on Sound Recordings ix

Prologue: Catchers of the Living 1
FRAGMENT I. SAMBA DIALLO: "THE WAR OF THE WHITES" / "CATCHER OF THE LIVING"

Introduction: Listening to Acoustic Fragments 11
FRAGMENT II. JÁMAFÁDA: "THE WAR IS HORRIBLE"

1 Abdoulaye Niang: Voice, Race, and the Suspension of Communication in Linguistic Recordings 23
FRAGMENT III. ASMANI BEN AHMAD: "ONCE UPON A TIME"

2 Mohamed Nur: Traces in Archives, Linguistic Texts, and Museums in Germany 66
FRAGMENT IV. JOSEF NTWANUMBI: "WE ARE INITIATES"

3 Albert Kudjabo and Stephan Bischoff: Mysterious Sounds, Opaque Languages, and Otherworldly Voices 101
FRAGMENT V. MAMADOU GREGOIRE: "THE SEA REQUESTS FISH FROM THE RIVERS"

Afterword: Knowing by Ear 147

Acknowledgments 157
Notes 161
References 183
Index 201

Note on Sound Recordings

The original sound recordings discussed in this volume, most of which are held by the Lautarchiv of Humboldt University in Berlin, can be accessed in person at the Lautarchiv using the inventory numbers provided. Placing the recordings online is a sensitive matter, because the speakers who were recorded in German POW and internment camps did not give their permission for the use and circulation of these recordings, although in many cases they did make attempts to communicate beyond the camps and thus did expect to be heard. Currently the content of many of the recordings remains unknown or untranslated, which makes it difficult to decide which of the recordings can or should be made public. We expect that these questions will be discussed during the process of restitution.

As of 2021, the Lautarchiv is housed at the Humboldt Forum in Berlin. For more information about the Lautarchiv, see https://www.lautarchiv.hu-berlin.de/en/sound-archive/.

Prologue

CATCHERS OF THE LIVING

> We relate, know, think, world, and tell stories through other stories and with other stories, worlds, knowledges, thinkings, yearnings. —DONNA HARAWAY, *Staying with the Trouble: Making Kin in the Chthulucene*

Theirs was an orderly arrival. They did not cross the Mediterranean Sea on overloaded, barely seaworthy, rubber boats. Then, thousands of young African men had been recruited or conscripted to fight in the armies of the Triple Entente during World War I. In France they were called *la force noire*.[1] In Germany, African men in particular were met with racist propaganda. As soldiers in French and British armies, however, they were not hindered from reaching European shores. Many had been forcibly enlisted to fight against the Central Powers (Germany and Austria-Hungary) in theaters of war in Europe. Some had come as *tirailleurs* (infantrymen from the French colonies), like the *tirailleur sénégalais* Abdoulaye Niang. Yet their experiences in Europe, their perspectives on this war, their accounts, stories, or narrative translations of what they saw have rarely surfaced in the colonial archive.[2] In other words: their trace is faint, predominantly visual, and formed by racist and racializing practices of representation in Europe (Diallo and Senghor 2021, 3).

For the phonographic recordings of German linguists, the Wolof speaker Abdoulaye Niang sang of France's policies of enlisting Senegalese men, who were then sent to join forces in Europe. For Jámafáda, a soldier from Dahomey (now Burkina Faso) who spoke Mòoré for the recordings that were produced in German camps, this was "the war of the whites." To him, as he relates on one of the acoustic recordings held at the Berlin Lautarchiv, the war was horrible and meaningless. He had been unable to find his brothers, who had been enlisted earlier, but was forced to march on. From South Africa, so-called war workers were sent to support the British army as members of the African Labor Battalions. On their way to Europe, six hundred war workers from South Africa drowned in the icy waters of the Baltic Sea when the SS *Mendi* sank off the Isle of Wight on February 21, 1917. The survivors took their narratives of the disaster back to South Africa. Reminiscences of these versions of the historical event later appeared in songs. The following song was performed for the recordings of the musicologist Hugh Tracey by the Reitz Bantu Choir. The lyrics were published in isiZulu and English in Tracey's book *Lalela Zulu: 100 Zulu Lyrics* in 1948; the composer's name is given as Siyiyo; the translator is not mentioned:

Iqanawe iMendi yathatheka khona e'wandle
Yashona iMendi namadodana ase Africa.
Kawufanekise ulwandle
Kawufanekise inqhanawe nabantu!
Washona uMendi, whashona uMendi,
Washona olwandle
Washiya inkedama
Washona uMendi.
Nawe manzi usibingelele
Xolani bantu bonke
Nithinina maAfrika?
Musani ukulala obuthongo.

The ship "Mendi" went down at sea
And sank there with the sons of Africa.
Can you picture the ship with the people in it?
Down went the "Mendi"
Down into the sea.
Many were the orphans that were left,
With the sinking of the "Mendi"
We fear you, waters of the sea
Soften your hearts, you people.
What do you say, Africans?
Stay not asleep below! (10)

Elements of these narratives and songs that traveled back to South Africa with the survivors still circulated in the 1980s within the musical genre of *isicath-*

amiya in Johannesburg (Erlmann 1995, 137).[3] Perhaps versions of Jámafáda's narratives traveled back to his country with him, too; perhaps they were shared and circulated. Abdoulaye Niang did not live to speak of his experience in Dakar. Jámafáda's and Abdoulaye Niang's recordings survived as ossified language examples in the Berlin Lautarchiv, together with hundreds of other acoustic echoes of prisoners and internees from World War I.[4] In Germany their spoken and sung texts were not considered historical sources until very recently. As language examples that were no longer of interest to German linguists, most of the recordings of the Lautarchiv were not translated for close to a century. Critical engagement with these recordings and the mode of their production was initiated by Britta Lange more than a decade ago. Her monograph on the history and the recordings of prisoners of war (POWs) in the Lautarchiv was published in 2020 (see also Hilden 2022; Hoffmann and Mnyaka 2015).

Yet, in 2018, an exhibition on the Lautarchiv at the Humboldt Box—the pop-up museum that advertised the anticipated collections and the fiercely criticized concept of the Humboldt Forum in Berlin, then under construction[5]—once again told the history of the opportunistic operation of recording linguistic samples with POWs as a legacy of both pioneering academics and the invention of the phonograph. Read with Ursula Le Guin and via Donna Haraway, this kind of historiography can be understood as a particular genre, a heroic tale, or "prick tale," as Haraway calls it, in which everybody apart from the heroic linguist(s) becomes "props, ground, . . . or prey" (Haraway 2016, 118; Le Guin 2019). For a long time, the speakers were omitted not only as actors from the historiographies of the war but also as contributors to the history of the Berlin Lautarchiv that keeps their acoustic trace.[6] The speakers were not presented as heroes, the begetters of the story of this peculiar archive; nor were their stories deemed anything but language examples, the catch of colonial linguists. Apart from ignoring hundreds of spoken and sung texts that multiply and version the narratives of World War I, this tale has systematically erased the history of epistemic violence in colonial linguistics and musicology, not only with regard to the Lautarchiv. It also continues to absent particular historical sources that testify to the presence of African migrants in Germany and to their part in German history of the first decades of the twentieth century.[7] Apart from their echo in the Berlin Lautarchiv, many of the speakers also left traces in other archives. These visual and textual records were distributed according to discrete areas of interest—to art museums, photographic collections, and anthropometric collections—or were published as linguistic texts. Only by means of reassembling this evidence, that is, by listening and reading across a variety of archives and disciplines, do traces of particular speakers surface in the present (Harrison, Byrne, and Clarke 2013; Lowe 2015; Hoffmann 2023).

Thus, while the Lautarchiv holds the echo of their presence, it is but one archive within a larger network of archives, depositories, and institutions in different countries that hold traces of African soldiers, war workers, and other African men who got caught in Germany in the turmoil of World War I. Examples of the materials preserved include several paintings and drawings representing the Muslim scholar Mohamed Nur from Somalia, as well as the biographic details of his life that appear in a grammar of the Somali language (see chapter 2). The grammar did not appear under Nur's name. The paintings, created by German artists, refer to him by a pseudonym he may or may not have chosen. Body measurements and anthropometric descriptions of the Congolese-Belgian intellectual Paul Panda Farnana and of Abdoulaye Niang, found in an archive in Vienna, can be reconnected to Abdoulaye Niang's recordings in the Lautarchiv (see chapter 1). Other sources for the writing of this book are recordings of drumming by the Congolese-Belgian soldier Albert Kudjabo, which were aired on German radio in 1924; and correspondence on the character of the Togolese migrant Stephan Bischoff that survived in the Hamburger Staatsarchiv (see chapter 3). Another example is a small card from the registers of the Red Cross that identifies Josef Ntwanumbi, a merchant seaman from South Africa, as a civilian internee held in the Engländerlager (English camp) in Ruhleben (see figure 2.3; his name is given as Twanumbee).

Like the acoustic recordings, most of the written or visual traces are part of the debris of imperial knowledge production. Many of the images of African soldiers and migrants who got caught in the war belong to the realm of autopsy—in the sense of *auto-opsis*, or seeing with one's own eyes—a practice that constructed racial difference as visual evidence (Weninger 1927; Doegen 1941; Berner 2003; Lange 2013; Evans 2010b). Some images are the result of voyeurism. A postcard that shows Paul Panda Farnana in soldier's garb, standing at the door of a train stopped at a German station with a group of white people below, staring at him, is an example.[8] The centenary of World War I prompted the (re)circulation of these racializing images. Again, the individuals subjected to this gaze, like Paul Panda Farnana during his journey through Germany on the way to an internment camp, are rarely identified; they almost always remain unnamed, are often beautified, and mostly racialized.[9] The archival results and residues of practices of examination, visualization, and representations of foreigners in Germany, in which propaganda, exoticism, and imperial science are interwoven, do not permit retroactive unraveling. All the traces I reassembled in this book are in some way connected to the zealous project of recording all the languages spoken by POWs in German internment camps during World War I.

The collection of acoustic recordings of POWs held at the Berlin Lautarchiv, which is the starting point of this book, was initiated not to conserve the histories of African soldiers, or of prisoners, or of civilian internees of World War I. Nor did German linguists of the Königlich-Preußische Phonographische Kommission (Royal Prussian Phonographic Commission; KPPK), which operated from 1915 to 1918, aim to document the prisoners' experiences of being interned in German camps while they awaited the end of the war. Despite the KPPK's limited focus on languages, the acoustic traces at the Berlin Lautarchiv are unique: they accumulate to a polyphone echo that resounds with the presence of prisoners and internees from all over the world. This echo prompted me to listen closely, to listen together with those who were able to understand and to translate the words spoken or sung, to learn to attend to the traces of the prisoners' presence by ear, and to begin to assemble the splinters of these traces in other archives.

Some of the one hundred recordings in nine African languages that were translated for my project by Phindezwa Mnyaka, Serigne Matar Niang, Fatou Cissé Kane, Johannes Ossey, Gilbert Katanabo Muhito, Faustin Sambu Avetsu, Constance Kutsch Lojenga, Dishon Kweya, and Bodhari Warsame[10] allow one to read and listen to echoes of the Great War as historical sources that speak of experiences of imprisonment. They present subaltern enunciative positions and refer to colonial histories in what are now the Democratic Republic of Congo (DRC), Somalia, and Ghana. They also present unexpected comments on and critique of colonial politics or the violent practices of evangelization. The recordings I selected for this book are not all that is to be found at the Lautarchiv, as the work of Britta Lange (2020), Irene Hilden (2022), and Ignazio Macchiarelli and Emilio Tamburini (2018) shows. The recordings with speakers of African languages are part of a linguistic survey that also holds recordings from Asia and Europe.

My research on recordings with African prisoners began with a set of recordings in isiXhosa, which Britta Lange, acting as a temporary custodian of the Lautarchiv, had given me in 2011 to take to South Africa (where I was working at the time). The translations by historian Phindezwa Mnyaka, as well as several sessions of collective listening in workshops at the Archive and Public Culture Research Initiative at the University of Cape Town in 2012, made clear once again that paying close attention to musical and textual genres is crucial for the project of making sense of these historical sound recordings. In particular, the apparent impossibility of finding out more about the singer Josef Ntwanumbi (spelled Twanumbee in the files of the Lautarchiv) initiated my long-term research into the Lautarchiv recordings with African POWs. The striking depersonalization of the prisoners in many of their traces in relation to the deafening propaganda that had dehumanized them seemed to call for more detailed research. This book is the result of following their available audiovisual traces over many years, not only to find out more about their journeys but also to learn what can be known by ear, whether and how acoustic sources speak differently in comparison with written sources, and what it means to listen to historical records in search of colonial history.[11] This research reflects my interest in the biographies of African prisoners and migrants in Germany during World War I, their journeys, and the narratives and songs they recorded. My earlier and ongoing work on sound archives, on orature from and in southern Africa, and on colonial history in relation to epistemic practices has provided a starting point from which to make sense of specific genres of orature and from which to unravel its metaphoric content (see Hoffmann 2009b, 2012, 2023).

As historical echoes of the presence of African soldiers and migrants in Europe in the first decades of the twentieth century, the Lautarchiv sources bear the trace of journeys from Aden on the Arabian peninsula, then British-ruled, to Germany with a *Völkerschau* (ethnological exhibit); they include narratives that speak of the experiences of colonial soldiers and poems that refer to the situation in Ituri, then part of the Belgian Congo, in the first decade of the twentieth century; or they comment on an event of German colonial violence against Yewe priests and a shrine in Ghana in 1913. In particular, the recordings with Mohamed Nur, Stephan Bischoff, and Albert Kudjabo provide fragments of a long history of migration to Europe, of the exploitation of African men as exotic bodies, and of their deployment as informants and also as teachers of African languages at the Seminar für Kolonialsprachen at the Hamburgisches Kolonialinstitut (Seminar for Colonial Languages at the Hamburg Colonial Institute).

I propose understanding the Lautarchiv as an oratory, in the sense of an accumulation of orature in the form of fragmented, polyphonic echoes of a war and the operations of colonial linguistics. These polyphonic echoes consist of orally transmitted stories, songs, narratives, and prayers. Some of these traveled from elsewhere or spoke of, or to, places and realities outside the camps and to topics that were unrelated to the aims of the linguists. The acoustic fragments accumulate into kaleidoscopic echoes of colonial histories, which open the possibility for other historical narratives and testify to other per-

P.1 Hermann Struck, *Portrait of Samba Diallo*, 1916. From Struck 1917, plate 92.

spectives (Pandey [2000] 2012). So far, historical voice recordings with non-European speakers appear in German museums mostly as ambient sound, or as acoustic wallpaper, gesturing toward an unspecified "exotic" world. In an attempt to present aspects of the polyphonic echo of POWs in a German museum as an assemblage of voiced utterances, I created the exhibition *Der Krieg und die Grammatik: Ton- und Bildspuren aus dem Kolonialarchiv* (War and grammar: Audiovisual traces from the colonial archive). This exhibition at the Museum am Rothenbaum—Kulturen und Künste der Welt (MARKK), Hamburg, on view from October 2019 to February 2020, framed the audiovisual trace of the Somali intellectual Mohamed Nur in a chorus of recordings of other internees and prisoners, but also made audible some recordings of the German Kaiser and the philologist Wilhelm Doegen (see plate 16).[12]

As a curatorial practice, this project translated into weaving together voice recordings from the Lautarchiv to form a new constellation of enunciations that then spoke to the opportunistic practice of knowledge production in German internment camps. In the sound installation in the first room of my exhibition, Jámafáda's and Abdoulaye Niang's recordings responded to the declaration of war by Kaiser Wilhelm II, which is listed as recording number one in the Lautarchiv. Yet the Kaiser, whose personal funds financed the operation of the KPPK, was not recorded in a camp. Nor does putting Jámafáda, Albert Kudjabo, and the German Kaiser into conversation undo what many of the prisoners and internees experienced in Germany. However, as a curatorial strategy of sympoesis, engaging with the recordings of the Lautarchiv follows Donna Haraway's suggestion to "stay with the trouble" of violent heritage (2016, 125). This allows their polyphonic trace to speak, as I will show in this book, beyond the violence of colonial knowledge production.

Fragment I

Samba Diallo: "The war of the whites" / "Catcher of the living"
TRANSLATED FROM BAMANANKAN BY ANKE NEHRIG

Jeneba Nahawa juru lasumaya
Ko n ye wara dugu la, Nahwa
. . . saya tɛna jɔn to, Nahawa
Sayajuru tɛna jɔn to.
Ah kelɛma ɲuman dɔn
tubabu kelɛ
kelɛ ma ɲuman dɔn, ɲenamaminɛ,
kelɛ ma tulɔnkɛ.
. . . n ye wa la diarraforo la,
kɛntiwara bɛ mɔgɔ dumu na, ɲenamaminɛ

Junuba Nahava, play the strings slowly
When I was away, Nahava
Death does not spare anyone
The strings of death do not spare anyone
The war knows nothing good
The war of the whites
It knows nothing good

Catcher of the living
The war is no game
I came to the lion's field
The lion eats the men
Catcher of the living

(Berlin Phonogramm-Archiv, Phon. Komm. 113)

In the collections of the Berlin Lautarchiv and Phonogramm-Archiv, there are at least two, perhaps three, speakers documented with the name Samba Diallo, although sometimes it is written as "Sambadialo." The name is very common in West Africa. The singer of the song quoted here was recorded by the musicologist Georg Schünemann, in 1916. Diallo is identified as being thirty-three years of age and as coming from Bougounie—then French Sudan (now Mali). Apart from later being recorded again, this time by Wilhelm Doegen, Diallo was also measured, examined, and photographed by Rudolf Pöch and Josef Weninger. In Weninger's 1927 publication, *Eine morphologisch-anthropologische Studie*, Diallo appears as "Samba Dschalo," and as "no. 35." He is presented as a Kautschuk trader who was married and had two children. Most likely he is the subject of a portrait by Hermann Struck (see figure P.1). He is not the speaker with the lip injury called "Mamadou Sambadialo," who appears in chapter 1.

Introduction

LISTENING TO ACOUSTIC FRAGMENTS

> As I began developing parts out of pieces, I found that I preferred them unconnected—to be related but not to touch—to circle but not to line up, because the story of this prayer was the story of a shattered, fractured perception resulting from a shattered, splintered life.
> —TONI MORRISON, "The Writer Before the Page"

> *Fragmentation*—that maverick which breaks into Clio's estate from time to time, stalls a plot in its drive to a denouement and scatters its parts. —RANAJIT GUHA, "Chandra's Death"

The English term *fragment* is related to the Latin *fragmentum*: a remnant, something always already broken, or broken off from a larger entity. The related verb *frangere*, to break, gestures toward a force but also toward imperfection, to a notion of the incomplete, of detachment (as in breaking *off*), indicating missing parts and the irreversibility of loss but also, inevitably, to imagination and suspense, perhaps even desire. Toni Morrison states that, in her writing, she has given preference to fragments instead of giving in to the desire for the "whole thing," like waking up from a dream and wanting to remember "all of it, although the fragment we are remembering may be—very probably is—the most important piece in the dream" (2019, 66). For historiographies, the question of whether the fragments we have found in the archives are the significant parts of a longer narrative, a discursive formation, is often hard to decide. The fragments that surface in an archive can have the character of a maverick, as Ranajit Guha describes in his essay

"Chandra's Death," one of the foundational texts of subaltern studies (Guha 1987). Fragments may speak, like sherds of pottery, from another time, yet they often do not allow the weaving of a "complete" story. In the case of the Berlin Lautarchiv, acoustic fragments are held together by an overall plot: the opportunistic practices of knowledge production in POW and internment camps during World War I. Yet as a cohesive element for the actual recorded narratives, this plot is weak. It is a porous container that holds spoken narratives and songs, often only until one listens closely, or until the recordings are translated and yield meaning, which the archive so far has omitted.

Josef Ntwanumbi ("Twanumbee" in the archive records),[1] a merchant seaman from the Eastern Cape in South Africa, was interned as a foreign national in the Engländerlager (English camp) in Ruhleben, near Berlin. His voice recordings were produced at the Odeon recording studios in Berlin, in 1917. On one recording he compared his experience of captivity with the time of seclusion he lived through during the rites that accompany male circumcision, perhaps intimating a sense of déjà vu (see Fragment IV). In March 1917, Albert Kudjabo, a Congolese soldier in the Belgian army, was asked to drum on a Melanesian "speaking drum" that had been borrowed from the Museum für Völkerkunde und Vorgeschichte (Museum for Ethnography and Prehistory) in Hamburg and was brought to the POW camp in Soltau for this purpose (see plate 11); he also sang several songs (see chapter 3). In April 1917, the Somali civilian internee Mohamed Nur recorded fragments of a poetic, polyphonic debate on the Dervish movement in his country (see chapter 2). Also in Ruhleben, Stephan Bischoff presented a fable that criticized the European evangelizing mission in Africa.

These speech acts, songs, requests, enunciations, remarks, comments, stories, prayers, and pleas, held at the Berlin Lautarchiv, are part of the debris of colonial knowledge production bequeathed to us (the inhabitants of the present) in the form of the colonial archive. For this book, I take the colonial archive at large as an imbrication of the discursive sense of "archive"—which determines what can be said and what was said, written, published, filed, and has become knowledge (of various kinds) as part of "imperial formations" (Stoler 2016)—together with "archives," as specific sites, collections, and institutions. The colonial archive, in this tentative definition, is based on the paradigms and epistemic constellations at work in exploring, describing, visualizing, and inventorying subjugated territories, resources, and people. This includes not only the research on a given people's languages and their music but also the creation of racial fantasies based on the examination of their bodies (Weninger 1927). The colonial archive is predicated on, shaped by, and thus intrinsically connected to "imperial formations," their power relations, and their agendas (Stoler 2016). Both the colonial archive in general and specific archives, as depositories of collections of documents and materialized knowledge that have been created with and for the colonial project, actively direct the work of researchers studying colonial history (Lalu 2009; Lowe 2015; Trouillot 1995).

Colonial history, in this sense, is not merely the history of former colonies but the history of everything touched by imperial formations. The question of how to create meaningful historiographies in the present, historiographies that can escape the epistemic frameworks created by discourses and documentation based on coloniality, has seen much scrutiny and debate in the last decades (Hamilton et al. 2002; Hamilton 2013). And while the interest in audiovisual collections has grown, too, the debates around the colonial archive have, to my knowledge, rarely included acoustic collections (Hoffmann 2020a). Yet historical voice recordings may allow for an understanding of subaltern speaking positions from within projects of imperial knowledge production, as well as beyond this immediate situation.

On the now-digitized sound files of the Berlin Lautarchiv, one hears spoken lists of words, repeated syllables, examples of counting, example sentences, and instances of "free speech" (in the grammatical sense). Sometimes these recordings are interrupted by a cough, by muffled laugher, or marked by a sense of unease in a voice. Many of these recordings transmit preserved aspects of repertoires, ossified by their very configuration as examples of languages. Once listened to as carriers of meaningful words, and/or as texts, as performative utterances and not as linguistic samples, these sound recordings from the colonial archive may allow for the surfacing of disturbing splinters of conversations, constrained by the power relations of the projects that produced them. The belated attention to their existence as carriers of meaning beyond their function as linguistic specimens—for instance, as in the case of the hundreds of recordings with colonial prisoners of World War I at the Lautarchiv that are presented in this book—may lead one to expect a retrieval of "stories untold." Yet what one can hear now often leaves listeners with bewildering fragments—echoes from the debris of imperial knowledge production. With regard to their semantic content, the archival composition of narratives, songs, counting, and all kinds of speech acts in one collection at the Lautarchiv, or in the much bigger Phonogramm-Archiv at the former Ethnological Museum of Berlin (now part of the Humboldt Forum), often seems arbitrary. The reason for this is a mix of language barriers, disciplinary paradigms, and disinterest in anything but the desired object—representative samples, or specimens, of music or *langue* (not *parole*).

A typical example is the situation in which the linguists at the POW camp at Wünsdorf, who neither understood the languages they recorded nor regarded the moment of recording as a dialogue, acoustically documented and filed a soldier's urgent plea not to be deported to another camp (see my further discussion of this Lautarchiv recording, PK 1114/2, in chapter 1). The plea was registered in the written files as a "narrative in Wolof." The soldier's appeal did not prompt or necessitate a response from the linguist at the time. The recording remained untranslated for a century, yet it was studied in 1943 with a focus on the racial features of voice (Bose 1943/44). With regard to this and other recordings in the colonial archive, the epistemological and archival configuration is remarkably durable. And in this way, the obliteration of the soldier's plea was carried over to the twenty-first century, into the now-digitized files, which demonstrates that digitization does not magically rescue acoustic files from obscurity.

Although I have sometimes suspected intentional dissimulation, mostly there is no cause for the regular and flagrant disinterest and subsequent archival omission of the semantic meaning of these historical recordings other than the configurations shaped by colonial epistemologies and contemporary practices of archiving.[2] Neither does archiving a plea as a language example count as an error in the archival records. Instead, in the light of the rationale for the recordings of the Königlich-Preussische Phonographische Kommission (Royal Prussian Phonographic Commission, KPPK), this makes perfect sense. The KPPK, a group of philologists, linguists, musicologist, and anthropologists, had set out to record all languages spoken in the internment camps in Germany. The semantic content of the resulting recordings, the spoken text they preserved, was deemed irrelevant for the production of linguistic records. The textual content of the recordings was thus not taken into account in the ordering, registering, and archiving that these samples underwent in the process of their itemization.[3] This means that the archival configuration that still exists in the present is the direct consequence of the specific epistemological approach undertaken by the original recordists. This written registration in many cases produced a permanent, yet not always irreversible, distortion of the semantic content of the acoustic objects. As a result of this process, it is impossible to retrieve a specific recorded utterance on the basis of the colonial register. The plea of the distressed *tirailleur*, for example, is archived as a "narrative" (*Erzählung*). However, the very nature of acoustic recordings, such as those with African prisoners of war, allows recorded, spoken words to be available, despite having been epistemically configured as linguistic objects and often wrongly labeled. The recordings themselves, due to their existence in that particular medium, have not been overwritten by the transcriptions and interpretations of researchers.

The linguists' disinterest in the semantics of the performed speech acts was neither unusual nor specific to this particular project. While manuals for the phonographic "collecting" of music and languages had already been published in Germany from the late nineteenth century onward, the strategic move to release these recordings from their disciplinary sequestration does not tell listeners in the present *how* to listen to them (Ankermann 1914; Sarreiter 2012; von Luschan 1896). Although the manuals requested particular practices of record keeping—for instance, with regard to the place and date of the recording, the person recorded, the genre, and language—and although the manuals often requested the transcription of the spoken or sung text, these details are rarely reflected in the written record. In many cases the documentation of the recordings is sketchy, if not absent. And even with collections for which meticulous documentation has been delivered and filed—as in the case of the recordings produced by the KPPK in World War I internment camps—this does not mean the documentation will tell listeners in the present what they desire to know. Often, the unexpected, voiced interventions come without explanation. Prompted to speak into the phonograph, speakers often told stories, gave accounts, sang songs, or recited poems that were part of larger repertoires or archives that had traveled to Germany with the soldiers and migrants.

In the sense of their textual, discursive content, these recordings can be described as acoustic fragments because, like the textual snippets Guha has studied, they often deny the denouement of a situation or episode we witness as listeners. This fragmentary nature is reflected in the appearance of specific recordings, their translation, and the often-sparse information on their speakers, as illustrated in this book by Fragments I–V. These Fragments are short insertions between the chapters. They present texts about which and speakers about whom I was not able to find more than what the written files of the Lautarchiv documented, aside from, in one case, a snippet from the Red Cross documentation issued to POWs in Germany; or in another case, a drawing created by the artist Hermann Struck; or in yet another case, comments on a recording or speaker penned by a member of the KPPK. These insertions signal an echo of a speaker that does not tell an entire story, yet they gesture to the speaker's historical presence with a trace even more faint than those of the speakers who appear in the chapters of this book.

The circumstances of the creation of these recordings, together with the at times enigmatic content of recorded speech acts, resonate with Michel Foucault's description of archival snippets he happened to stumble upon in the archives: those vignettes that were "brief, incisive, often enigmatic" and that threw a spotlight on the "lives of infamous men," or that, perhaps, made them "infamous" in the first place (Foucault 1967, 161). There are, however, major dissimilarities: the acoustic fragments—which are the object of our close listening and the starting point for my project of reassembling archival traces—were not produced to document or address disobedience, the trespassing of laws, or violations of the social mores of a time; nor do they speak to contested definitions of sanity. Few contain life stories or biographies, which became the signature of later attempts to create oral archives of social history. Instead, the recordings selected for this book were produced in the attempt to account for, systematize, and describe the grammar, phonetics, and lexicography of different languages. Additionally, although all speakers were registered in personal files in the written archival records, and sometimes their images or imprints (in photographs, casts, or body measurements) were also entered into the archives of anthropometrical documentation, on the recordings, one hears people speak and sing (mostly) in their own voices. These "voices" one hears are *recorded* voices; they are produced, not collected (Sterne 2003). They are *mediated* echoes of voices (Hoffmann 2023). *Voice* in this case does not refer to the political voice, that omnipresent metaphor.

Particularly in orally transmitted texts, the voice of the speaker is often not congruent with the political voice of a subject (as in "having a voice"). Yet, intentionally or not, even as a recording, the human voice carries meaning beyond words, which is an effect that may evade the neutralization, or reduction, of the acoustic trace in linguistics (Schrödl and Kolesch 2018). Voice generates an excess of meaning that cannot be reduced to that which can be transcribed (Dolar 2006; Weidman 2015). Understanding voice as a phenomenon on the threshold of *aesthesis* and *logos*, as always gesturing *and* speaking, and thus carrying both semantics and an affective quality, as Doris Kolesch and Sybille Krämer (2006, 7) write, presents voice as a vehicle of speech that imbues it with a moment of unruliness or potential ambiguity (see chapters 1 and 3). Still, the presence of a recorded voice does not necessarily signify subjectivity. Narrative agency is that which may well speak beyond or outside the meaning of a speech act that was recorded as a linguistic example, yet it is often intrinsically linked to a polyphonic source of an oral repertoire. Therefore, I suggest understanding these voice recordings as echoes that reverberate with what was once said or sung, yet that still carry meaning beyond text, even if the speakers remain unknown, and the unknown source of an echo may disturb the listener's sense of direction.

Although every acoustic fragment is shaped and mediated, often effaced and deformed by the practices that led to its capture, not every fragment speaks to the epistemic practices directly: whereas some of the recordings with POWs may speak of imprisonment, of recruitment politics, of the war, and of the ones left behind and missed, other speakers leave aside this situative frame of reference and tell a story that takes flight from the misery of imprisonment, the trenches, the biting cold of German winters, the drama of the Great War. In this way, the acoustic fragment may speak from a position beyond the set of practices that led to its production as a recording, complicating the identification of the frame and referring to a repertoire or discourse that has not been, and could not have been, captured in its entirety (see chapters 2 and 3).

The project of translating about one hundred of these recordings in African languages intensified my understanding of the arbitrariness of their compilation in the Lautarchiv, where they are kept as acoustic fragments that transmit resonances, echoes, of which the source (the speaker) may or may not be known. Reassembling the traces of the speakers, as well as writing this book, was held together by the practice of close listening, together with translators who did not hear these spoken texts as language examples but instead attended closely to the genres and performativity of those spoken and sung texts. Listening, reconnecting, and translating (or retranslating) — but also the discussions that accompanied the process of interpretation — became a strategy for reactivating these acoustic fragments. Apart from learning more about genres of orature, this process has complicated my understanding of voice. A striking example of the shift of meaning that came with close listening can be found in the recordings on which So, a West African deity, spoke through Ruhleben internee Stephan Bischoff. These particular recordings demonstrate that close listening, translating, and reassembling acoustic fragments with their contexts from outside the archive may destabilize Western notions of evidence and the indexicality of voice. The recordings in which a deity speaks from a colonial archive also present an echo in which the sound archive overlaps with what Toyin Falola (2016) has called ritual archives, which may hold incantations and invocations and, thus, spiritual aspects of African histories that Falola does not expect to be present in the colonial archive.

The recorded echoes of voices I attend to in this book transmit the repercussions of content that float in a sea of acoustic traces, from remote places at times—if we take the metropolitan sound archive as the core of the endeavor—or from the center, Berlin. They were recorded for reasons that have little to do with the discourses, repertoires, or oral archives that resonate in and through them. While acoustic fragments indelibly bear the watermark of the power that generated them—the productive power of the archive, or the field of imperial knowledge production—they are not entirely created by the processes that prompted and conserved them (Stoler 2009). Although the initiator of the project now housed at the Lautarchiv, Wilhelm Doegen, saw himself as the creator of the recordings, attentive listening makes clear that the recording technique has *not* created the genres, the poetry, or the songs that were archived. It itemized them and stabilized them and thus, in a way, snatched them from the ephemeral quality of voice into the realm of concrete, collectible, sounding objects fixated in a potentially endless loop of repeatability. Recording as archiving has, in this way, arrested the versioning practices of orature, which have, for instance, actively altered and reshaped the narrative snippets that recall the sinking of the SS *Mendi* in 1917 (see the prologue). As acoustic recordings, sequestered in a German archive, these pieces of orature have been separated from the circuits of retheorizing that filter, reimagine, and alter them, over time, in the flow of oral performances and by means of musicking.

While voice recordings in the archive are part of the debris of colonial knowledge production, they are also fragments, or components, of larger entities: they may be splinters of the fabric of a discursive field, elements of a repertoire of songs and stories. The term *repertoire* here does not gesture toward sequentiality—as in an earlier, "primitive" version of an archive—rather, as Diana Taylor (2023, 22) suggests, these repertoires exist synchronously to practices that entail writing; they are not the lesser sisters of an all-encompassing archive, nor do they generally constitute an antihegemonic challenge to the archive as the locus of the "writing culture" of power. Repertoires, although they do contain nonverbal practices—such as in dance, performance, and a range of culturally informed and ingrained gestures—are not the antipode to the archive as a repository of text and images. Much of my engagement with voice recordings is preoccupied with words and texts in recorded, performative forms. Much of what was documented as examples of musical expression, or as samples of linguistic research material, is indeed also text, albeit never exclusively. Related to or coming from genres that belong to a specific repertoire of orature, acoustic fragments may encapsulate elements, themes, and topics that can also be found in written texts or in collections of artworks and photographs. As part of a repertoire, or as an element of a discourse, I suggest understanding the semantic content of many acoustic fragments as belonging to a hive, a flock of utterances that were once the interconnected (tightly or loosely) yet flexible parts of a formation. It was the claw of power of which Foucault speaks—here, that of colonial knowledge production—directed by the will to know "native languages" and to systematize the languages of the world (as seen from the colonial center)—that fished these fragments from the midst of an oscillating swarm of utterances and severed them from it. In the archive these fragments have ossified; but at times a glimmer of what once was a (re)sounding swarm remains. The historical recordings of the Lautarchiv preserve abbreviated and medially formatted pieces of orature, songs, or narratives. These are acoustic snapshots of a specific moment of voicing and performance. Yet, all recordings are produced—not collected. This means that although they may have initially belonged to a discursive formation, they were not taken away from orature or from the repertoire of a group of speakers or a location. The materialization of the acoustic snapshot thus did not deprive the repertoire of a vital element. The swarm of utterances that makes a discursive formation (orally or in writing) did not lose a component.

Today, together with thousands of other recordings in other archives and collections, the recordings of the Berlin Lautarchiv constitute a massive collection of resonating relics that can be heard as echoes of the presence of African men in Germany during World War I. This book follows some of their acoustic traces to other archives, along the paths of historical narratives, through networks of researchers, to the imbrication of their research on languages and concepts of race in German linguistics. By means of reassembling acoustic, visual, and written traces with close listening—which was often collective listening (Hilden 2021) together with the work of translation—the single narrative often told of the Lautarchiv becomes invalid: it is not the history of white men and their apparatuses, nor is it another story of pioneering practices of research and archiving. The project of my research was to read acoustic traces of the presence of African soldiers of World War I in Germany as aspects of colonial history that predominantly surface in the recorded orature in this particular German archive. These echoes often begin to speak clearly only in connection with other traces the speakers have left in different archives. In this way, the journeys of the speakers surface in the interstices between recordings, along archival networks, and in the shadows of other practices of research that had, by then, already learned to parasitize on war and colonization.

The chapters in this book are arranged around speakers as historical persons. Each chapter follows the traces its speakers' recordings have laid

Lfd. Nr _____

PERSONAL=BOGEN

Lautliche Aufnahme Nr: *P.K. 1116* Ort: *Berlin*
                                    Datum: *8/VI 17*
                                    Zeitangabe: *12 Uhr 20 Min*
Dauer der Aufnahme: ____ Durchmesser der Platte: *27 cm*
Raum der Aufnahme:
Art der Aufnahme (Sprechaufnahme, Gesangsaufnahme,
Choraufnahme, Instrumentenaufnahme, Orchesteraufnahme): *1. Zahlen 1 bis 58*
2. Erzählung.
                                    Mosi (Sudansp?)

Name (in der Muttersprache geschrieben):
Name (lateinisch geschrieben): *Jámafáda*
Vorname:
Wann geboren (oder ungefähres Alter)? *vielleicht 21.*
Wo geboren (Heimat)? *Batumno*
Welche größere Stadt liegt in der Nähe des Geburtsortes? *Fatägurma*
Kanton — Kreis (Ujedz):
Departement — Gouvernement (Gubernija) — Grafschaft (County): *Mosi*
Wo gelebt in den ersten 6 Jahren? *Batumno*
Wo gelebt vom 7. bis 20. Lebensjahr?
Was für Schulbildung? —
Wo die Schule besucht? —
Wo gelebt vom 20. Lebensjahr? *seit 1914 Deutschl.*
Aus welchem Ort (Ort und Kreis angeben) stammt der Vater? *Batumno*
Aus welchem Ort (Ort und Kreis angeben) stammt die Mutter? *Batumno*
Welchem Volksstamm angehörig? *Mosi*
Welche Sprache als Muttersprache? *Mosi*
Welche Sprachen spricht er außerdem? *Mosi, ein wenig französ.*
Kann er lesen? ____ Welche Sprachen? ____
Kann er schreiben? ____ Welche Sprachen? ____
Spielt er ein im Lager vorhandenes Instrument aus der Heimat? ____
Singt oder spielt er modern europäische Musikweisen? ____
Religion: *Heide* Beruf: *Soldat.*
Vorgeschlagen von: 1. *Meinhof.*
                   2. *Wildvogel.*

                               1. Urteil des Fachmannes
                                  (des Assistenten): *gut*
Beschaffenheit der Stimme: *Meinhof.*

                               2. Urteil des Kommissars: *Dunklere*
Mittelstimme mit hinreichender
Sonorans. *Wildvogel.*

I.1 Personal file of Jámafáda. Berlin Lautarchiv.

out to other archives and museums, and each chapter seeks to connect its spoken and sung texts to historical events. Abdoulaye Niang, Mohamed Nur, Stephan Bischoff, and Albert Kudjabo lead us through the history of the Lautarchiv. Their audiovisual traces speak of war and colonialism, of the exploitation of their presence in science and art, but also of their navigation of their roles to their own ends; of genres of speaking, singing, and drumming; of their migration to Europe; and of the wish to leave it. These songs and spoken texts, which sometimes seem to, and perhaps actually did, comment on the recordings of others, present archival echoes that have remained faint and yet add to the kaleidoscopic trace of the presence of African soldiers and internees in Germany from 1911 to 1922.

Fragment II

Jámafáda: "The war is horrible"
TRANSLATED FROM MÒORÉ BY ANONYMOUS

Ma ba riki maa kissi nassara kiengue guiabre. Guiabre ka noomeye. Guiabre da kassog ye. Ma kiengue, ma ka yein maa koendama. Maa koendama nda kiengue guiabre. Ma baass sarwissi fada n'gourma. Ma kaa yein yee. Ma yok kiende, ma na kaa yein ma koendamye yii menguiye. Diilfaa, n'mii kuiyan mii kaa kui ma ka yein la. Ma yok kiende yok kiende ma ra ka yein. Guiabre daboosogye. Kuili yooma tan ma ka yein ma n'ma, ma ka yein ma n'ba. Ket nena, ka yein kasruima. Ma kietin kiena ring ka yein ye. Ma n'da yi yok kiende yooma tan ma ka yein ma poa ye ma ka yein biiye. Mii ma na beekay, ma kui ma ka kui. Guiabre san saa ma kuili, ma na kuili meyein. Guiabre ka saa ma kuikai saa.

My father has sent me to the Whites to go to war. But the war is dreadful. So I went, but I did not see my older brothers who had been recruited for the army already. I have not seen them there. I marched, but I did not meet them. So I ask myself whether they are dead or still alive, because I have not seen my older brothers. The war is horrible. Since three years I have not seen my mother and my father. So far I have not seen any one of my older brothers. I am still in this uncertain situation; I have no perspective. Since three years, since I have left I have not seen my wife and my child. I do not know whether I can weather this. If I could only survive and return to my home, if the war could end. If I could go home and see my family. If this war doesn't end and I die here, everything ends.

(Berlin Lautarchiv, PK 1116/2)

Jámafáda spoke Mòoré and came from Fada N'Gourma, which was then French Sudan (today it is Burkina Faso). In his personal file at the Lautarchiv (PK 1116), Wilhelm Doegen estimated Jámafáda's age as twenty-one and stated that he had been a soldier since 1914. Jámafáda's narrative was recorded in Wünsdorf, in the so-called Halbmondlager (Half Moon, or Crescent, camp) where mainly Muslim prisoners were interned. Yet Jámafáda was registered as heathen (*Heide*). After meeting Jámafáda in Wünsdorf, Carl Meinhof described him to Felix von Luschan as intelligent and remarked that his appearance could be of interest for the anthropologist. Doegen published his anthropometric photograph, which focuses on the scarification of his face (and is therefore not included here) with no indication of his name.

Jámafáda's account of his recruitment at his father's request and his search for his older brothers, who were also in the French army, was recorded as an example of Mòoré for the Lautarchiv. The account was filed as "eine Erzählung" (a narrative), with no indication of its content.

In 2010, Jámafáda's text appeared, probably translated for the first time, in the documentary film *Boulevard d' Ypres/Ieperlaan*, directed by Sarah Vanagt.[4] For my project, Jámafáda's narrative was retranslated by a Mòoré speaker who wishes to be anonymous due to his precarious situation as a refugee in Germany. In 2019, Jámafáda's voice recording interrupted a speech of the German Kaiser Wilhlem II in my exhibition *Der Krieg und die Grammatik: Ton- und Bildspuren aus dem Kolonialarchiv* (War and grammar: Audiovisual traces from the colonial archive), at the MARKK, Hamburg (see plate 16). In 2020, his narrative was published in Britta Lange's book *Gefangene Stimmen*, which later appeared in translation as *Captured Voices* (2022).

1

Abdoulaye Niang

VOICE, RACE, AND THE SUSPENSION
OF COMMUNICATION IN LINGUISTIC
RECORDINGS

> *Et nous voilà pris dans le rets, livrés à la barbarie des civilisés.* (And here we are trapped in the nets, left to the barbarity of civilized men.) —LÉOPOLD SÉDAR SENGHOR, "Au Guélowar" (Camp d'Amiens, September 1940)

Abdoulaye Niang's voice sounds clear; the record barely crackles. He sings of Senegalese men who are sent to the abattoir; he was one of them. His recordings speak from within the colonial archive, yet his songs come from elsewhere. They resound with echoes of another place and time, with speaking positions absent from the written documents of the Berlin Lautarchiv and absented again in a recent presentation of this archive in the provisional exhibition space called the Humboldt Box, in Berlin.[1] On the now-digitized recordings with the file numbers PK 1114–15 at the Berlin Lautarchiv, one can hear Abdoulaye Niang speak and sing in Wolof. The slaughterhouse he sings of was the Great War, in which he, a *tirailleur sénégalais*—member of a French infantry unit drawn from colonial Senegal—had fought until he was captured and interned in a German POW camp.

In the archive, spoken words coagulate, sometimes to the point of incomprehensibility. Stored as recordings, sequestered in particular collections, and rarely translated, these words, sentences, and texts have been severed

1.1 Abdoulaye Niang, undated photograph. From Doegen 1925, 18.

from vital discursive constellations, from repertoires, and from ways of speaking or singing. Despite a growing interest in the collections of the Lautarchiv in recent years, most of what the POWs sang and spoke into the gramophone of the Königlich-Preußische Phonographische Kommission (KPPK) in the years 1915–18 has not been used for the crafting of historiographies of World War I. With the exception of some recordings that were repatriated in recent years, most of the recordings of the Lautarchiv are not known in the countries where the singers came from.

This chapter follows the voice recordings the KPPK produced with Abdoulaye Niang in 1917, which have become his archival echo, the conserved and distorted resonance of his presence in Germany. Abdoulaye Niang's spoken words lay out an acoustic track that leads to archives in Germany and Austria. His recorded words and voice allow me to revisit the double inscription in the sound recordings: the ways in which they speak as recordings, and what can be known of their modes of production. This double inscription, which is often found in historical voice recordings, speaks, on the one hand, to the practices of recording and archiving. On the other hand, the recordings with Abdoulaye Niang have conserved instances of enunciation that testify to his attempt to communicate. Engaging with historical voice recordings and their documentation that together tell of moments of knowledge production, of situations of speaking, of the raison d'être of an archive, I seek to keep intertwined in my writing what was irreversibly interlaced and concomitant with the act of recording from the start: the practice of exploiting the presence of POWs with the aim to create a collection of voice recordings, which was the foundational moment of the Berlin Lautarchiv, together with what was sung and told by those POWs yet rarely given the status of meaningful utterances. The sound recordings resonate with the prisoners' encounters with the KPPK. The recordings are the echoes through which these moments of knowledge production surface in the present; they speak of and beyond the moment of recording.

The writing of an interface as an interface, that is, of historical sound recordings as objects that contain layered, incommensurable constellations of practices, does not always make for a fluent narrative. The story has to follow the curves and hooks in the logic of colonial knowledge production and the politics of archiving, while maintaining a grip on the concurrent ways in which strategies of speech, aspects of discourses, and biographic snippets surface both acoustically and in writing. Thus, this first chapter does not seek to create a sequential arrangement of an entangled situation. Rather, it retraces words that have disappeared and resurfaced; the chapter follows Abdoulaye Niang's archival traces in Europe.

In this book, acoustic archival traces are not the often-romanticized, faint remnants that disturb the researcher's sensibilities but never seem to allow for a systematic reading of (in this case) the practices of colonial knowledge production (Stoler 2016, 5). In the acoustic traces that have survived at the Berlin Lautarchiv, the agendas of the speakers surface, and the coloniality of specific epistemological and archival practices has left its mark. The recordings bear fragments of subaltern presences in speaking—often sotto voce, yet at times quite boldly—some of which came wrapped in specific genres, and some of which were aspects of discursive formations that froze and fell silent in the archive.

Abdoulaye Niang's voice was recorded in the studios of the Odeon record company in Berlin, around noon on December 8, 1917, by the linguist Carl Meinhof and the philologist Wilhelm Doegen.[2] These acoustic recordings produced a century ago are part of a collection of around 2,700 linguistic and musical recordings compiled by linguists and musicologists of the KPPK in German internment camps from 1915 to 1918.[3] This venture was the first systematic project on this scale in Germany. The KPPK's aim was to phonographically record languages and musical expressions to create an archive of acoustic specimens. As with other collections of recordings, these acoustic records were produced to conserve ephemeral sound, thereby amassing a body of comparable specimens for research in linguistics and musicology (Sterne 2003; Brady 1999; Hoffmann 2023, 2020a, 2020b, 2021a; Lange 2013, 2020). This means that, unlike objects that were selected to be filed and archived *after* their production, such as the correspondence between researchers, or the collections of photographs from an expedition, the POW recordings at the Lautarchiv and the Phonogramm-Archiv were recorded with the explicit intention of creating specimens for an archival collection.

All nine records produced with Niang were archived as language examples of Wolof, a language he came to represent, although he spoke Wolof with a Lebou accent.[4] The recordings produced with Abdoulaye Niang were translated for the first time in 2013 by Serigne Matar Niang, who was able to identify his accent.[5] Serigne Matar Niang's translation of and engagement with the spoken and sung texts of his namesake revived the lyrics of what must have been a contemporary song, perhaps one circulating in Dakar at the time Abdoulaye Niang enlisted in or was recruited for the French army.

In the written documentation of the Lautarchiv, one song is listed by Wilhelm Doegen, who did not understand Wolof, as "Lied mit Händeklatschen" (song with clapping). The song speaks from the position of a wife watching her husband being sent to the European battlefields. Years after he had left Dakar and had fought in the war, Niang sang this song for the linguistic record. His voice animates aspects of a local debate about French military recruitment campaigns—a debate that had crossed the ocean, traveled with the soldiers to the theaters of war in Europe, and then found its way into the camps. With its translation, the short song surfaces as a fragment, severed from what may have been a communicating swarm of utterances in popular culture, with poetic statements and responses that engaged with the recruitment of thousands of Senegalese men who fought for the French army in a war that was not theirs. Serigne Matar Niang's transcription and translation of the song recording with Abdoulaye Niang (Berlin Lautarchiv, PK 1115/3) reads as follows:

> Saa dieuker djoundaw dafa oblise dji baay (han . . .).
> Gnii nga betoir
> Gni nga dieum Farance, gnithi dess moblisedji baay (han . . .)
> Saa dieuker djoundaw dafa oblise dji baay (han . . .).
> Gnii nga betoir—
> Gni dieum Maroc, gni nga dieum Farance, gnithi dess moblisedji.

> My young husband has been conscripted to be sent to the abattoir.
> He has been bound
> These are sent to slaughterhouse
> My sweet husband is heading toward France
> These are sent to the abattoir—some to Morocco and France and the others are on standby.

In Europe, there is little to read, let alone hear, of the war Samba Diallo (Fragment I) and Abdoulaye Niang sing about. Only in recent years has the involvement of colonial soldiers in the war become a more regular part of the historiographies of World War I.[6] The aim to decenter, to multiply, and to version the historiographies of World War I, as articulated by Santanu Das and others, is not easily achieved.[7] As an example, compared with the mass of autobiographical texts on the Great War written by Europeans (and Americans), very few published accounts of African soldiers exist. There are only a handful of memoirs by West African veterans of World War I, one of which was written by the Senegalese griot Bakary Diallo, who attained French citizenship after the war.[8] The absence of letters, poems, and memoirs, which Das (2011b, 6) describes as the "cornerstones of European war memories," on the side of the troops from the colonies may not only be the result of what has been often assumed to be the illiterate background of many of the colonial soldiers. Arguably, the absence of these documents may be a result of, for instance, African soldiers preferring other forms of representation and cultural mnemonic traditions. In other words, autobiographies may not have been their chosen form for articulating war experiences. Neither does the absence of letters from public archives necessarily connote their nonexistence. The lack of written sources, likewise, does not confirm an absence of memories, which may still circulate in the more elusive medium of orature, in songs and storytelling. These orally transmitted memories are hard to trace and are less likely to be held in European archives than are written letters, poems, or memoirs.

The war that Samba Diallo presented as "Catcher of the living" (Fragment I) rarely attracted nationalist sentiments on the part of African soldiers in the French army. In Abdoulaye Niang's song (above), a young wife describes this war as an abattoir, a place of certain, violent death. On another recording (Berlin Lautarchiv, PK 1115/2), Abdoulaye Niang sings of people who pick up cigarette butts in the streets:

Borom megot dotoulene cigar
Rene nguene deme
Maroc bay

Dear people who are picking up butts in the streets
You won't have a chance to smoke cigars!
This year you are going to Morocco![9]

In this song, which repeats the three lines cited here several times, Niang seems to refer to those who were conscripted as men who had lived in precarious economic situations, men whom the war would cut off from their prospects and hopes. To understand songs in which the war is depicted as an abattoir, as well as Diallo's notion of the war as a "catcher of the living," it is of some importance to note that a large percentage of the estimated 135,000 West Africans who were sent to the European battlefields were conscripted. Moreover, French authorities, writes Myron Eschenberg, "made little effort to transmit an ideology to Black African soldiers." Neither were attempts made "to communicate the aims of this war to the men who were conscripted or recruited" (1991, 365; see also Njung 2016, 2020).[10] The politics of the annual conscription quota in West Africa, the responsibility for which was handed down to local chiefs, was met with suspicion, with tactics of avoidance and resistance (Lunn 1987, 29; Page 1987, 5; Quinn 1987, 108). In Senegal alone, writes Christian Koller (2017), 35,000 men fled from French recruiters (see also Senghor 2021).

Samba Diallo's song and Abdoulaye Niang's recordings in Wolof are elements of a many-voiced body of recordings at the Lautarchiv that have survived as acoustic documents of a survey (Hoffmann 2014a). I hear these recordings as echoes of World War I but also as fragmented echoes of colonial histories and the presence of thousands of colonial soldiers in Europe in the first decades of the twentieth century. As with all echoes, it is difficult to locate a speaking position, and those who speak or sing often cannot be seen. What can be heard with a delay may be distorted, abbreviated, and opaque. One hundred years after the recording, the positionalities of the speakers cannot be fully known, yet the speakers were never only informants and rarely spoke as informants at all.

Only now, belatedly, and from the "seams of intelligibility" of this textual and sonic archive (Foucault 1981, 189), are we, the heirs of this archive, beginning to listen to these acoustic recordings. One could hear the recordings of the Lautarchiv as a kind of oratory, sung and spoken by POWs. Yet this is an oratory that the linguists did not plan to create, nor did the prisoners or inmates orchestrate it. The recordings were produced and filed as an enormous collection of specimens for linguistics and comparative musicology. They were not intended as expressive, performed texts, and, as such, they were buried for a century in the Lautarchiv of the Humboldt University in Berlin (earlier, the Lautabteilung of the Kaiser Wilhelm University).

The recording sessions were systematically arranged and scheduled by the KPPK, and the records on shellac disks (Lautarchiv) and on wax cylinders (Phonogramm-Archiv) were produced following a protocol of documentation, prefigured with regard to their form, yet much less so with regard to their content. Standardized protocols of acoustic and written archiving were followed, with the aim of (further) establishing the new technology of phonographic documentation as an objective, scientific practice. This entailed one invariable questionnaire, the *Personalbogen* (personal file), filled in for each speaker or singer (see figure 1.2). The mode of documentation probably followed the instructions for phonographic recording developed at the

Lfd. Nr 1114.

PERSONAL=BOGEN

Lautliche Aufnahme Nr.: PK/1114 Ort: Berlin (Wünsdorf)
Datum: 8/ XII 1917
Zeitangabe: 11 Uhr 30 Min
Dauer der Aufnahme: 2½ Min   Durchmesser der Platte: 27 cm
Raum der Aufnahme: Aufnahmeraum der Oswa Werke
Art der Aufnahme (Sprechaufnahme, Gesangsaufnahme, Choraufnahme, Instrumentenaufnahme, Orchesteraufnahme): 1. Zahlen 1 bis 40.
2. 2 Erzählungen.

Wolof (Sudaneger.)

Name (in der Muttersprache geschrieben): [Arabic] Abdulaye
Name (lateinisch geschrieben): Nyang Abdulaye
Vorname:
Wann geboren (oder ungefähres Alter)? 39
Wo geboren (Heimat)? Jore (Insel)
Welche größere Stadt liegt in der Nähe des Geburtsortes? Dakar
Kanton — Kreis (Ujedz):
Departement — Gouvernement (Gubernija) — Grafschaft (County): Senegambien
Wo gelebt in den ersten 6 Jahren? Jore
Wo gelebt vom 7. bis 20. Lebensjahr? Dakar
Was für Schulbildung? arabische Schule in Dakar, auch französische Schule
Wo die Schule besucht? Dakar
Wo gelebt vom 20. Lebensjahr? Dakar
Aus welchem Ort (Ort und Kreis angeben) stammt der Vater? Jore
Aus welchem Ort (Ort und Kreis angeben) stammt die Mutter? Jore
Welchem Volksstamm angehörig? Wolof
Welche Sprache als Muttersprache? Wolof
Welche Sprachen spricht er außerdem? Wolof, Bambara, Ful, Suon, Französisch, englisch
Kann er lesen? ja   Welche Sprachen? arabisch, französisch
Kann er schreiben? ja   Welche Sprachen? arabisch
Spielt er ein im Lager vorhandenes Instrument aus der Heimat? nein
Singt oder spielt er moderne europäische Musikweisen? nein
Religion: mohammedanisch Beruf: Kaufmann
Vorgeschlagen von: 1. Meinhof.
2. Wilh. Doegen.

Beschaffenheit der Stimme:
1. Urteil des Fachmannes (des Assistenten): gut.
C. Meinhof.
2. Urteil des Kommissars: Kräftige helle Stimme mit gutem Klangansatz.
Wilh. Doegen.

1.2 Personal file of Abdoulaye Niang. Berlin Lautarchiv.

Vienna Phonogrammarchiv, which was founded in 1899 (Lange 2013, 338).[11] Predefined bodily postures, such as a specific angle of the head, and a uniform measure of distance to the gramophone made for a regulated choreography of bodies in the recording context. The questionnaire, Britta Lange (2013) suggests, was thought to allow for the later identification of the origin of the speakers of the dis-located, recorded speech acts in the archive. Scientific as they were deemed to be, these protocols and guidelines did not always work in the actual recording situation. While the men with whom the recordists of the KPPK could communicate were asked to tell specific stories or pronounce series of words, the speakers and singers with whom they could not converse spoke more freely. As a result, particularly speakers of non-European languages became orators, or griots, of war, of captivity, and of colonial histories they were never asked to chronicle or comment upon.

In Colonial Prisons, on Show Grounds, and in Camps: Practices of Knowledge Production

The identification of POWs in Germany as an opportunity to record foreign languages did not occur in a vacuum. In the wake of already established practices of exploiting subordinated peoples for the production of knowledge, German researchers saw the internment of around 650,000 POWs in 1915 as a serendipitous opportunity for the ambitious project of recording all the languages present in the camps. Already before the war, the main initiator of the project, the philologist Wilhelm Doegen (1877–1967), had envisaged a "museum of voices" (*Stimmenmuseum*). After the war, in his "Denkschrift über die Errichtung eines 'Deutschen Lautamtes' in Berlin" (1918; an unpublished manuscript arguing for the establishment of a "sound bureau"), he wrote:

> The sound collection created by the Kommission [in the camps] provides new material for the research on languages and for the exploration of foreign countries . . . recordings from distant and farthest (*fernste*) countries promise extraordinary valuable scientific gains (*Ausbeute*).[12]

Doegen further stressed the importance of the recordings for the teaching of languages as "dienlich für koloniale Interessen und Ziele" (instrumental for colonial interests and aims). What exactly these aims and interests were, he did not specify. From 1909, the philologist, together with the Odeon record company, had produced a series of publications: *Doegens Unterrichtshefte für die selbstständige Erlernung fremder Sprachen mit Hilfe der Lautschrift und der Sprechmaschine* (Doegen's instructions for the autonomous acquisition of foreign languages with the help of the talking machine). The *Unterrichtshefte* were published together with records of literary classics in French and English, spoken by French and English actors. Encouraged by the success of his acoustic teaching material, Doegen applied for the creation of a Königlich-Preußisches Phonetisches Institut (Royal Prussian Phonetic Institute) with the Prussian Ministry of Education in 1914. As a result, the Königlich Preußische Phonographische Kommission (KPPK) was established in late October 1915. The aim of the KPPK was to produce linguistic and musical recordings in German internment camps. The operation was financed by the Kaiser's personal funds (*Kaiserlicher Dispositionsfond*). Doegen was not, as he later would claim, the director of this Kommission. Instead, as a member of the KPPK, he was responsible for producing the actual acoustic recordings on gramophone records in the POW camps. Under the chairmanship of Carl Stumpf, the director of the Berlin Phonogramm-Archiv, the KPPK assembled around thirty researchers, among whom were the anthropologist Felix von Luschan, the linguist and Africanist Carl Meinhof, and the musicologist Georg Schünemann. Carl Meinhof was responsible for the recordings in African languages. His subordinate collaborators were Martin Heepe, whose interest lay in languages from Madagascar and Comoros; the anthropologist Paul Hambruch; and the linguist Otto Dempwolff. Along with the 1,650 spoken language recordings that the KPPK produced in the POW camps during the war, more than one thousand wax cylinders with musical recordings were produced. On Doegen's request, the musical and linguistic recordings were later divided according to discipline: the musical recordings were, and are still, held at the Berlin Phonogramm-Archiv; the linguistic recordings are kept at the Berlin Lautarchiv.[13]

In addition to the documentation at the Lautarchiv, correspondence among the KPPK members tells of its operations and aims:

> In which camp are the Benin people? I have not seen them. In Wünsdorf there is a Baule—Ivory Coast—with four pointed teeth, from my point of view a typical Sudan negro. He can drum nicely. There are also good Ful (Toucouleur) and one Wolof and a fine Comoro man.[14]

Carl Meinhof sent this inquiry to the anthropologist Felix von Luschan in July 1915. Together with his question to von Luschan of "how best to get access to the coloreds in the Muslim Camp," this passage from the correspondence of two leading figures in the KPPK speaks of epistemological practices and of a readily appreciated opportunity to get hold of captured men.[15] It also documents Carl Meinhof's racializing practices and interests. The results of the operation of the KPPK later appeared in a number of publications by the Lautabteilung (Lautbibliothek 1929a, 1929b; Lautbibliothek and Westermann 1936), which held the sound collections after the war, as well in publications by Wilhelm Doegen (1925, 1941), Martin Heepe (1920), and Carl Mcinhof (1939). Doegen also referenced the operation in his unpublished pamphlet of 1918.

The KPPK's project of anthropometric, anthropological, and linguistic research in Germany was not singular: in a similar project in Austria, prisoners of war were targeted for research, too. Unlike the project of the KPPK in Germany, the research on POWs in Austria focused mainly on physical anthropology and racial studies. The Austrian project was initiated in 1915 by the Anthropologische Gesellschaft (Anthropological Society) in Vienna (Lange 2013; Berner 2004) and was led by the anthropologist Rudolf Pöch. Both projects have received increased scholarly attention in recent years. Britta Lange, in particular, has delivered very detailed work on the methodology, epistemological practices, and preconceptions underlying the recording and anthropometric research on prisoners of war in Germany and Austria.[16]

While the procedures of the KPPK (and its Austrian twin) have attracted scholarly interest in recent years, this is rarely the case for smaller, or less well-documented, projects of acoustic recording, which were also often combined with anthropometric examination of the speakers and singers (Garcia 2017; Kalibani 2021; Hoffmann 2023, 2009a). Sound archives like the Berlin Phonogramm-Archiv (founded in 1900) and the Vienna Phonogrammarchiv (founded in 1899) regularly commissioned acoustic recordings, yet these were mostly of minor import for the researchers who delivered them.

By 1915, attempts to record the languages and musical expressions of non-European people had been exploiting the systematic subjugation of people in the colonized countries for some time, and research projects had also benefited from colonial wars (see, for instance, Berner, Hoffmann and Lange 2011). In Berlin, attempts to accumulate a substantial collection of sound recordings for the Phonogramm-Archiv's research in comparative musicology were based on opportunistic practices from the start. In the early years of the Berlin Phonogramm-Archiv, its directors regularly visited the sites and performers of *Völkerschauen* (ethnic shows) in the region to produce musical recordings. Susanne Ziegler's (2006) catalog of the wax cylinder recordings of the Berlin Phonogramm-Archiv lists thirty-six collections of musical recordings that were produced under such circumstances. Some of these collections were recorded with performers in zoos, including the collections Archiv Siam (1900), Archiv Samoa (1910), and Archiv Sudan (1909). The opportunity that musicologists saw in these shows also led to recordings at ethnic displays like the "Somali Dorf" (Somali Village) in Luna Park, Berlin (Archiv Somalia [1910]); at Castan's Panopticum, Berlin (Archiv Tunisia [1904]); and at Carl Hagenbeck's traveling shows (Archiv India [1902]). Another example of opportunistic practices of acoustic archiving is the collection of recordings made with Joseph Tjikaya from Congo, whom Ziegler lists as "Leo Frobenius's boy" (Archiv Kongo [1908]; see Ziegler 2006, 88).

Farther from home, European anthropologists and linguists favored colonial prisons and police stations, camps, and pass offices as locations where colonized people could be coerced into being recorded, examined, photographed, and sometimes cast in plaster (Hoffmann 2009, 2020b; Lange 2013). While phonographic recordings, in contrast to the "acquisition of ethnographic objects," did not extract the songs and stories from the speakers and singers, projects of phonographic recording were informed by a colonial attitude, which guided and justified the epistemic practices and routines of racist anthropology.[17] These violent practices of acquiring objects and extracting information, together with the anthropometric examinations to which European researchers subjected people of color in camps, prisons, and *Völkerschauen*, did not merely calibrate methodologies and shape emerging fields, like comparative musicology and *Afrikanistik* (African studies); they also created a horizon of expectation on the part of the researchers.[18]

Some of the members of the KPPK had prior practical experience with coercive methods of colonial knowledge production. For example, a decade before he became a member of the KPPK, Felix von Luschan had measured, acoustically recorded, photographed, and made casts of people in pass offices, police stations, and prisons in South Africa (von Luschan 1906, 863). Von Luschan's correspondence with Lieutenant Zürn of the *Schutztruppe* (Germany's colonial military force in Africa) during and after the colonial war in German South West Africa (now Namibia) testifies to von Luschan's unscrupulous exploitation of the situation of subjugated people to acquire human remains for his collection (Zimmermann 2001, 244). The Austrian anthropologist Rudolf Pöch, who would visit the POW camp in Wünsdorf, Germany, on von Luschan's invitation in 1917, had also made maximal use of the disempowerment of the people he classified as "Bushmen" during the colonial war in South West Africa and during a severe drought in the Kalahari. He coerced people into enduring anthropometric examinations that took up to six hours, and filmed, photographed, and acoustically recorded them in 1907 and 1908 (Hoffmann 2023). Pöch had also stolen human remains on a grand scale, using the logistical structure of the *Schutztruppe* to arrange his travel and to transport his loot to Austria (Hoffmann 2023; Schasiepen 2019; Rassool and Legassick 2000).[19] The majority of the historical collections produced under these circumstances are still owned by European institutions.

In recent years, linguists have started to review the coloniality of their discipline and are finding that the exploitation of subjugated and disempowered colonized people was a common practice in colonial linguistics (Errington 2008; Deumert, Storch, and Shepherd 2020). The German philologist Wilhelm Bleek produced linguistic research and documentation of languages with inmates of the Breakwater Prison in Cape Town. The prisoners were released from prison for this purpose and subsequently confined to Bleek's private house in Mowbray, a suburb, in the 1870s. On this particular "opportunity" made possible by colonial politics, which included the disappropriation, eviction, disempowerment, and genocide of local populations in South Africa, Carl Meinhof (1913, 6) wrote: "Sad as this [the criminalization of their way of life and subsequent imprisonment] was for the Bushmen, for science it was a blessing . . . in this way Bleek could study Bushman languages with his friends ǁKabbo and Han'kasso to his heart's content, and create invaluable collections"(see also Bleek 1868; Bank 2006). For his *Polyglotta Africana* of 1854, the missionary linguist Sigismund Koelle had collected samples of one hundred African languages in camps for freed, formerly enslaved people in Sierra Leone (Irvine 2008). The World War I–era project of recording and examining people in German POW camps, with a particular interest in people of color, was thus a continuation of colonial knowledge production in proximity to, and profiteering from, war, crisis, and the subjugation of people.

Thus, what may now sound like a particularly bizarre convergence of war and linguistics followed a well-established tradition of colonial knowledge production on an augmented scale. When the "catcher of the living" ensnared thousands of foreign men in Germany, this was welcomed by German linguists and anthropologists as an exciting opportunity. Rudolf Pöch would describe the situation enthusiastically as "eine ungeahnte Völkerschau" (an unexpected ethnic show; see Lange 2013, 75; Pöch 1916). Leo Frobenius published his propagandistic book *Der Völkerzirkus unserer Feinde* (The ethnic circus of our enemies) in which he berated the Triple Entente as a *Rassendompteur* (a tamer of wild races) in 1916.[20] This would not hinder Frobenius from doing research in the camps and becoming a camp overseer in Romania in 1917, where POWs had to excavate prehistoric graves (referred to as *tumuli*) for his research (Kuba 2015, 108; Frobenius and Freytag-Lohrinhoven 1924; Weule 1915).

Wilhelm Doegen published his populist anthology *Unter fremden Völkern: Eine neue Völkerkunde* (Among foreign peoples: A new ethnology), in which the camps are described as a "field site" of ethnology, in 1925. He also cited Kaiser Wilhelm II's remark, "Dann haben wir die ganzen Kerle für Jahrtausende auf Platte!" (Then we will have all the chaps on record for centuries!; Lange 2015, 85). Propagandistic discourses and notions of trophies of war were thus not clearly distinguishable from what was seen as scientific research with imprisoned men from the armies of the Triple Entente.

Carl Meinhof's racialized list of a "typical Sudan negro," "a fine Comoro man," and a "good Ful" shows that men of color became specimens in the operation of the KPPK—not only for the establishment of an acoustic collection of the languages of the world but also as examples of and for anthropometric research into the alleged interrelation of language and race. In October 1915, Meinhof wrote to Felix von Luschan, "Hopefully there will be an opportunity to point out that our linguistic research must be complemented with your anthropological research."[21] There are other examples of the imbrication of categories of language and race that predate the massive uptake of racial ideologies in linguistics in Germany during the 1930s (Römer 1984; Meinhof 1912). One example of research that sought to bolster this constructed connection between language and race is a 1913 publication on the examination of larynges in which the author searched for but was unable to demonstrate any physical differences between Africans and Europeans (Grabert 1913). Another research example is Wilhelm Bleek's idea that clicks in southern African languages constituted the missing link between humans and apes (Bleek 1868). Pöch repeated Bleek's assertion in his diaries, as did the Polish linguist Roman Stopa in the 1930s (Stopa 1935).[22] Wilhelm Doegen's brother, who was a dentist, produced so-called *Palatogramme* (casts of the mouth) with prisoners in the camps for the KPPK; these casts are now missing. It is not clear whether they were used for research purposes.[23]

That Is the Part (You Throw Away)

The operation of the KPPK in German camps was meant to be secret, yet the undertaking of the acoustic recordings and the interests of the linguists are well documented in the massive (unpublished) paper trail of documents housed at the Berlin Lautarchiv.[24] Sometimes remarks, written in the margins, reveal the researchers' attitudes or refer to their networks. In a letter to all members of the Kommission, chairman Carl Stumpf explained the secrecy of the operation in the camps by referring to the Ministry of War's concerns.[25] At the same time he made clear that, in academic publications, the project of recording could be mentioned:

> Niemand wird etwas daran finden das deutsche Gelehrte die Gelegenheit benützen durch den Verkehr mit den Gefangenen ihre Kenntnis fremder Dialekte zu erweitern.[26]
>
> Nobody will find it problematic that German scholars use the opportunity to expand their knowledge of foreign dialects by means of communication with prisoners.

The attempted secrecy of the operation allowed Wallace Ellison, the author of an autobiographical novel about his internment at Ruhleben, in the Engländerlager (English camp; a camp for foreign nationals), to dismiss the (fictionalized) inmate Toby Robert's angry outcry, "If you gramophone me, I'll kill you!" as sheer paranoia (Ellison 1918). Yet it was not the secrecy of the operation during the war that absented what some of the prisoners said and sang from the written archive.

What made some words and stories disappear from the written archive was neither secrecy nor an outright prohibition to speak of war and imprisonment during the recording sessions. The occlusion of words and semantic meaning from the written documentation was, rather, the effect of the "internal procedures" of an emerging field, which, in the form of predetermined protocols that classified, ordered, and regulated what was and was not relevant to the archiving of languages, had the effect of excluding some enunciations from being heard and recorded in writing.

As mentioned above, during the recording sessions, there were moments of "free" speech, that is, speech that was not regulated with a request for a specific text or genre. These instances of "free speech" were prompted by the researchers' request that the prisoners speak or tell a story into the gramophone funnel; however, what was said was often unintelligible to the recordists, who did not understand the languages spoken or sung. This means that unexpected spoken texts were archived in the Lautarchiv. These recordings may have been connected to, or may have been part of, practices of record keeping of a different order. The recordings with Mohamed Nur, for example, who chose to cite a passage from Somali orature that speaks of local histories, are an example of this phenomenon.[27] In many other recordings, the prisoners spoke or sang of the war. But because the operation of the KPPK was not conceived as an exercise in producing sound documents for the crafting of a social history of war, imprisonment, or recruitment, these spoken texts that could have been—and still can be—relevant for historiographies were marooned in the Lautarchiv as potential (but mostly unused) linguistic examples of unknown semantic content.

The belatedly realized possibility of hearing and reading the Lautarchiv as an interface, or a meeting place, of different practices of record keeping—written and orally performed archives but also a variety of discursive formations—is not merely the result of the language barriers that were in effect at the time of recording. Rather, it is the result of the researchers' preemptive dismissal of *parole*, in the sense of concrete, situated, semantically meaningful, and individual events of speech. Their disinterest led to the regular and durable failure to register certain narratives and enunciations in written form. Yet the technique of acoustic recording allowed for the survival and conservation of these spoken texts whose content was deemed irrelevant. While the researchers refused to hear and write *parole* in favor of *langue*—which was their designated object of study—it was in fact *parole* that the gramophone recorded.[28] In the realm of the colonial archive, this makes for an exceptional aperture: that which was not deemed relevant and therefore was not recorded in writing in some cases still exists as acoustic documents. This means that one can now revisit the recordings to listen to the speech acts and songs themselves instead of reading what is left of these voiced texts after they have been filtered through a written archive of colonial knowledge production.

Still, the occlusion of many of the spoken texts is durable: in the case of a great number of recordings, especially those in non-European languages, the catalog of the Lautarchiv obscures the topics of the spoken texts more than it informs about them. The persistence of this exclusion was reflected in the aforementioned exhibition on the Phonogramm-Archiv and Lautarchiv at the Humboldt Box in Berlin (2018). Even now, translations or transcriptions of the recordings scarcely have a part in the representation of the archive. In this recent exhibition, the Lautarchiv again appeared as an archive created by pioneers of linguistics, a collection of marvelous sounds (and not of meaningful words), and a demonstration of early recording technologies. Its representation at the Humboldt Box spoke of the work of prominent researchers—not of an archive that holds collections of spoken texts voiced by hundreds of prisoners of two world wars.

Another example of the longevity of absenting performed speech emerged in an exhibition on German colonialism at the Deutsches Historisches Museum (2016). A recording from the Lautarchiv spoken by the POW Massaud ben Mohammed ben Salah, a trader from the kingdom of Bornu, was included, accompanied by his portrait, painted by Hans Looschen (Frobenius and Freytag-Loringhoven 1924, figure 27).[29] However, as in so many cases, the voice recording was reduced to acoustic wallpaper at the exhibition by playing it as sound with no available transcription or translation. What the African POW had to say was thus turned into background noise. The practice of ignoring ben Salah's words is particularly disturbing because a historical transcription of his account in Kanuri is available as a publication of the Lautabteilung. The transcription makes it clear that ben Salah is speaking of his failed attempt to defect to the German army (Lautbibliothek 1929b).

For the larger project of collecting languages, the interests of the KPPK linguists meant that comparable recordings of vocabularies and standardized stories were the preferred objects. One such story was the biblical fable of the prodigal son, of which more than three hundred versions were recorded in at least twenty European languages and hundreds of dialects (mostly from the United Kingdom and France). The fable was a standard template for the creation of comparable linguistic specimens, used also, for instance, in George A. Grierson's *Linguistic Survey of India* (1903–28).[30] Obviously, the story (of the prodigal son) could be requested only from the soldiers who knew it. The comparability of languages, in this case, was confined to the reach of the literary canon of Christianity.

In all cases, what was said or sung in the moment of acoustic documentation was transformed or recoded into recordings as examples of grammar, syntax, phonology, and pronunciation and into lexicographical lists. There are few indications of interest in specific genres or practices of communication or linguistic codes, such as drum language. For the researchers, the focus on linguistic examples did not entail a categorical exclusion of the possibility of meaningful words being spoken during the recording of examples of languages. Yet, these were not deemed pertinent to the research. The result was a momentary, one-sided suspension of speech and song as communication, paired with, and exacerbated by, the recurring inability to understand non-European languages. As mentioned above, the recordists asked the prisoners to speak into the gramophone funnel to record words and sentences that were to be researched—at a later date—as examples of language. The recurring inability to communicate with the prisoners also shows in the incomplete personal files of many of the African speakers: questions could not be communicated, and answers were not understood (or both). Thus, many of the spoken texts were engraved into the wax of the records and entered into the archive as unknown, unaccounted-for texts. Whether and how the speakers read this suspension of communication is difficult to ascertain, and it may have varied for each speaker and in each situation. Yet the effect of this one-sided disinterest in communication was that spoken words fell into a chasm of meaninglessness in the Lautarchiv, from which some words did not surface for a century, and from which many have not emerged at all. Of the 450 recordings in forty-eight African languages, only very few have been translated to date.

Another consequence of the modalities at play in the recording situations was that the human voice lost its appellative efficacy. In the moment of recording, the language barrier, in combination with what can perhaps best be described as the magic of the discursive field, had the effect of turning the acoustic voices of many prisoners into mere sound, or phonetic signals (Bourdieu 1991; Hoffmann 2009a). Voice, as an ephemeral phenomenon in speech, in singing, and even in wordless chanting, exists between the physical body and a person's socialized environment; it is therefore malleable and of enormous, culturally specific plasticity. Yet the KPPK researchers reduced voice to sound. This becomes clear in the case of Abdoulaye Niang's voice being described as an acoustic phenomenon by a person who did not understand its meaning or appeal. The last section to be completed on the personal questionnaires of the Lautarchiv was "Beschaffenheit der Stimme" (properties, or characteristics, of the voice). This is one of the sections Doegen would always complete, despite being unable to understand what the African speakers had said or sung. What he described on most of the personal files is thus voice as an empty vehicle. Lacking any sense of the meaning and performative quality of voice, in the absence of tonal communication, Doegen described sound signals bereft of the human voice's ability to express, address, plead, or convince, if that even was what the speaker had tried to do. These signals were inscribed onto wax, converted to sound objects or (perhaps not only for the Kaiser) trophies, which were also indexical of the physical presence of captured enemies.

Yet the case of Abdoulaye Niang demonstrates that the human voice, as long as it sounds, cannot be neutralized. As an acoustic recording, voice retains an echo of its performative quality, even when listened to years later, during which conventions of speaking may have changed. This capacity of voice (even as a recording), together with the decolonial shift in the evaluation of the significance of acoustic documents, allows us (in the present) to revisit these acoustic documents as echoes of speech and of sonic traces of presence. Whether we can or cannot hear the registers of speech that were at play, or detect undertones a hundred years later and out of context, is another question. Thus, while the work of the KPPK generated and preserved an incommensurable discrepancy between the spoken, recorded words and the written documentation of linguistic examples, the recorded voices in the Lautarchiv can be revisited and listened to again from the "seams" of the archive.

Double Inscription and the Ornament of Registration

In the Lautarchiv the long row of *Personalbögen* amounts to a monotonous ornament of registration. Comparable to what Siegfried Kracauer has described as a (danced) "mass ornament," with performers who became "indissoluble girl clusters" (Kracauer 1995, 76), the Lautarchiv houses a mass paper ornament that is yet to be displayed as a long row of files that detach spoken or sung texts from speakers. This ornament of registration dissolved, rearranged, and de-individualized prisoners. Hundreds of these personal files document what may now be read as the systematic dissociation of the operation of the KPPK from the reality of the war that had brought the men into German POW camps to begin with. The files do not actively withhold the occasion of recording: after all, the location, *Kriegsgefangenenlager* (prisoner-of-war camp) is printed on about two-thirds of the files that were completed for each of the recordings.[31] Yet the camps—and with them war and imprisonment—appear as a location, not as the occasion they were, and certainly not as a topic or an integral element of the research and archiving practices they made possible. On every preprinted questionnaire, the recordist was asked to note the location of the recording session (usually the name of the camp), the number of the recording, the exact time of recording, and what was recorded (*Art der Aufnahme*). The singer or speaker was asked twenty-six questions, including the following: name, date of birth, place of birth, whether he went to school, whether he could read, occupation, birthplace of the speaker's mother and father, *Volksstamm* (tribe), religion, and so on.

These questions convey the interests of linguists and musicologists, which did not touch the subjects of war and imprisonment. In the questionnaire, only the inquiry as to whether the speaker/singer played a European instrument available in the camp relates to the situation of the prisoners. According to several hundred personal files in the Berlin Lautarchiv, the war that had captured the men before their voices were caught on wax or shellac was of no consequence for the linguistic and musical recordings of the KPPK. Nor did the possibility that representing "the languages of the world" with

BESONDERE BEMERKUNGEN:

Infolge eines Schrapnellschusses ein Stück des linken Oberlippe fortgeschossen.

1.3 "Special Remarks: Part of the left upper lip shot off by a shrapnel shell." Personal file of Mamadou Samba Diallo (or "Sambadialo") (detail). Berlin Lautarchiv.

a survey conducted *only* among young, captured men might be problematic occur to the (exclusively male) KPPK.

Perhaps it is the uniformity of the files, or the staleness of the repetitive absenting of a global catastrophe rendered irrelevant, that makes for the rupture when the actual effects of the war intrude on the quiet order of registration. An indication of its brutality punctuates the neat paper trail of acoustic recordings on the preprinted page attached to personal file number 289. The files with the numbers PK 286–89 chronicle the recordings produced with the *tirailleur* "Mamadou Sambadialo" (Mamadou Samba Diallo) on the afternoon of June 3, 1917. According to his personal file, the prisoner from West Africa spoke Yoruba, Bambara (Bamanankan), and what was described as *Dahomeen* (probably Fongbè). He was recorded in a barrack in the Halbmondlager in Wünsdorf, near Berlin. Under the rubric *besondere Bermerkungen* (special remarks) one reads: "Infolge eines Schrapnellschusses Stück der linken Oberlippe fortgeschossen" (Part of the left upper lip shot off by a shrapnel shell) (see figure 1.3).

This lesion had damaged the speech organs of the prisoner, who was of interest as an informant; it was therefore relevant to the recording of language. In the ornament of the written protocols, this comment on a battlefield wound registers as a minor perturbation that barely ripples the surface. Yet Diallo's injury was worth noting because it had the potential to interfere with the will to know, record, and archive as many languages as accurately as possible. Injuries that did not touch on a prisoner's ability to articulate words are absent from the forms. I learned of Abdoulaye Niang's wound on his left foot from another archive, for instance. In the acoustic recordings with Mamadou Samba Diallo, I cannot hear a speech impediment.

In this particular case, the order of documentation and what it reveals and conceals is reversed: here I cannot *hear* the repercussions of the war, yet they do surface in writing. The reversal of the scheme illustrates its logic: Doegen's special remark on the injury Diallo had suffered emphasizes the disturbing absence of the war in the grammar of archival registration that capitalized on, but kept quiet about, World War I. This scheme of regular elision of the conditions and modalities of knowledge production in the camps also shows in publications that draw on these archival records. In the grammars and articles of linguists that were published, sometimes decades later, and that, in several cases, formed the basis of academic careers, the preconditions of this massive survey are rarely mentioned.[32]

Disregarded on the preprinted questionnaires, the topic of war appears only randomly in specifications about what was recorded (*Art der Aufnahme*; type of recording), such as *Sprechaufnahme* or *Choraufnahme* (spoken recording or choir recording). About half of the narratives and songs in African languages documented in the written files give no indication at all of the content of the song or narrative. Among dozens of lists of words, recordings of counting, stories, and songs, very few recordings made with speakers of African languages are specified in writing as touching on war or imprisonment: a "war song" presented by Samuel Brown (*Kriegslied*, PK 863); the "song of a prisoner" by Ali ben Bedja (*Lied eines Gefangenen*, PK 868); a "narrative of war between Sayid Ali (?) and . . . ," by Ali Swahili (*Erzählung vom Krieg zwischen Sayid Ali und . . .*); and "an account of his life when the war broke out" by Makluf (*Aus seinem Leben bei Beginn des Krieges*, PK 1224) are strewn among many love songs and animal stories.

The translations for my project present performed texts that differ—often significantly—from the indications that surface in the registers of the archive in Berlin. The file that registers Abdoulaye Niang's *Lieder mit Händeklatschen* (songs with clapping) is an example. It does not itemize the content of his song, which presents a critique of French practices of military recruitment. The same goes for recorded performances listed as circumcision songs (*Beschneidungslieder*), presented by an isiXhosa speaker from South Africa, whose name was given as Josef Twanumbee (more accurately, Ntwanumbi; see Fragment IV). While Ntwanumbi did indeed sing of *abakhweta* (initiates), his songs artfully inserted experiences that speak of the camps into what were probably fragments of a song about boys' initiation (Lautbibliothek and Westermann 1936; Hoffmann and Mnyaka 2015). Asmani ben Ahmad's "account from his life" is actually the account of a journey that left him stranded in a German POW camp (*Erzählung aus seinem Leben*, PK 1108; Fragment III). Jámafáda's "narrative" (*Erzählung*, PK 1116; Fragment I) translates as his account of being sent to the "war of the whites." In other words, there is, in most cases, no information in the written register on the content, themes, or genres of the recorded songs and stories. The digitized register perpetuates this order of things: it allows searches only of languages (often misidentified) or of historical terms for ethnic groups, at least for the large number of recordings in non-European languages.

For listeners and readers who do not speak African languages, these accounts of war experiences surface after translation; the songs and spoken records speak of life in the camps: homesickness, loneliness, uncertainty, and fear. This makes for a clear divide: while the written files impart the interests of the linguists and the regulated modus operandi (or contention thereof), the spoken or sung records speak of other things. Thus in a semantic sense, the war surfaces in this archive of imperial knowledge mostly aurally: in reminiscences and laments, in stories of loss and sadness, in expressions of pain and fear, and in poignant articulations of the predicament of uncertainty about when the war will end. The recordings, with labels that identify speakers with misspelled names originating from often incorrectly specified places, were carefully stored, cataloged, and kept over the years. They were rarely used for further research. Listening to them now, together with translators—linguists, musicians, and historians—who understand the languages and to whom the performativity of the recorded voices is not lost, means listening to the recordings (also) as historical sources. Paying attention to content and genre emphasizes the striking incommensurability of written and acoustic files that is not simply a result of a divide between aural and written archives but an effect of the suspension of communication in favor of the collection specimens; it is a result of colonial epistemic practices, the denial of intersubjective knowledge production together with the dismissal of performativity and, at times, a language barrier. Close listening attends to an entirely different textual and sounding archive. The astounding contrast between the omission of the war in the written documentation and its poignant presence in the recordings makes for the impression of performative texts accidentally captured and subsequently buried in the archive that preserves them (Hoffmann 2023).

This chasm between the drama of war and captivity that appears in what was said or sung and the distorted or neutralizing descriptions that appear in the written documentation makes it seem easy to wrench the acoustic recordings from the order for which they were once created, and to unhinge them from the grip of colonial knowledge production to which they belong but which has not produced them faithfully as texts. The strategic move to recuperate the recordings as semantically meaningful, historical texts is absolutely necessary. This "change of the terms of engagement" with these spoken texts—the move from collecting and archiving linguistic examples (or from taking them as linguistic examples only) to listening to them as the spoken texts of imprisoned men—allows for the resurfacing of the excess of performativity (Shepherd 2015). Speaking, once understood, belatedly regains its appellative urgency for the listener. Listening closely may also direct attention to the agency that comes with specific genres that enables speakers/singers to enunciate in certain ways but that also allows speech to take flight, to leave the straightjacket of the requested language examples, and to transmit a message or express critique.

Listening to the recordings as meaningful texts retrieves the "recalcitrant presence" (Prakash 1994, 1489) of voiced texts as performances, of songs and accounts that speak of things that were of no interest for the linguists. It is important to note here that this recalcitrant presence of subaltern speaking positions emerges in historical recordings *despite* the fact that a subject position to speak from is often *not* granted to the "native informant" (Prakash 1994, 1488; Spivak 1988).[33] It is impossible to disconnect the acoustic recordings from the order of things that produced them, the disciplines that created them as specimens, and the archive that held them over the years (Foucault 1996; Hamilton et al. 2002). Listening to and reading these aspects of spoken texts attend to the complexities of an inextricably connected interface in which knowledge production, archiving, and subaltern speaking positions have met, and during which none of these speech acts had taken place or been recorded accidentally.

The recordists of the KPPK neither eavesdropped on nor recorded conversations incidentally—this was technically not possible with the recording technology available at the time. The precondition for each of the POW recordings of the Lautarchiv was that the person had been interned in a camp, happened to speak a language of interest to the linguists, and was therefore requested to speak into the funnel of the gramophone. This process of selection and subsequent request prompted each of the speech acts that was recorded.[34] The protocols of recording, for the sake of the acoustic quality of the recording, demanded that the speaker stand in front of the gramophone or phonograph and speak directly into the funnel.[35] For any historiography of colonial knowledge production and its "epistemologies of the acoustic," this is significant (Ochoa Gautier 2014). The *Personalbogen* that comes with each acoustic file reminds the listener that there was always a specific situation, a peculiar setting.

Sometimes what was said was in response to a request that does not appear in the files: for example, the demand to sing something in a specific genre, to speak specific syllables, or to present examples of a secret language.

Often these traces of particular interests and agendas have to be searched for elsewhere, such as in the case of the collection of recordings of drum language or in the case of recordings of a secret language from West Africa. In all cases, not merely the requests of the linguists but also the recording format mediated the speech act or song; the duration of time available on the physical recording disc or cylinder determined the length of the story told or the song presented.

While the formats of the recording discs or wax cylinders have mediated and often abbreviated what was presented by the speakers and singers, the specific constraints of the recording situation have also marked every song, story, or account. The demands or deformations of the recording situation, the requests of the recordists, the coloniality of the epistemologies at play, and even distortions that the medium itself caused cannot be removed. While the acoustic recordings may have left the grid of linguistic specimens performatively and semantically, they cannot be interpreted in a way that bypasses these indelible watermarks of archive (Stoler 2009, 8), the durable inscriptions of the will to know that watermark the recordings as part of the debris of colonial knowledge production.

Historical voice recordings differ crucially from archived written texts, such as those authored by colonial administrators, travelers, and ethnologists or those stemming from the minutes of trials, presenting transcriptions of what was said or may have been said, or exemplifying speech superimposed with someone else's writing. In the case of recorded speech, the acoustic signal of a voice has been engraved in wax or shellac. Unlike spoken testimony—to take on one of Ranajit Guha's now canonic interventions, "Chandra's Death" (1987), which was written by a "village scribe," translated more or less faithfully into scripture, and then converted into the generic conventions of the language of a testimony—archived voice recordings can be revisited as acoustically conserved, voiced text. Yet the existence of a conserved acoustic voice recording does not imply straightforward access to a subjective or political voice. In the best cases, one can hear the reverberations of the situation of speaking and recording. In many cases, depending on the quality of a recording, speech or songs can be translated (or retranslated). This means that instead of reading a historical transcription, making guesses about the bias of the writer, and speculating on his or her attitude toward the speaker and on what may have been left out (a cough, a sigh, a laugh, an alteration of voice), the listener can go back to a conserved speaking situation. Experts can be consulted to discuss the use of language and to attend to genres, to repertoires of stories, and to related narratives that resound in the recording. Perhaps even prosody, pitch, and modulation—the performativity of the voice—can be discussed and analyzed. It is precisely the double inscription of acoustic recordings—as objects of colonial knowledge production and also as recorded speech acts and songs once intonated by a human voice with performative abilities—that allows listeners in the present to know aspects of colonial histories by ear.

Unfortunately, but not accidentally, this double inscription is rarely transmitted in the archives' registers. Here too, both coloniality and disciplinary reason are built into the structure of archive: while digitization does a lot for the mobility of historical recordings—which can travel in this form, be listened to, or translated outside of the archive after listening collectively with speakers of the languages that were recorded—this does not magically alter or even "decolonize" the record. Misspelled names stay misspelled and misleading titles remain distorted after digitization. This means that the decolonial move to make these recordings available outside of the archive; to translate, hear, and read these recordings anew; and to listen to them together with speakers of the language spoken in the record as potential sources of colonial histories is rarely supported by what is contained in the archival registers.

While spoken or sung fragments may, in the moment of recording, have responded to the joke of another inmate, or may have imported repercussions of a larger discursive formation from elsewhere, many of these fragments have lost their context irreversibly to the effect of obscurity. Colonial epistemological practices prove to be durable (Stoler 2016; Hoffmann 2023). This underscores the longevity of the power relations and practical modalities of epistemological practices, the persistence of the order produced by the colonial archive. Some utterances from collections of acoustic archives can be revived and may achieve "retrospective significance" (Trouillot 1995, 58), while others are irreversibly petrified fragments that have lost all resonance. And while the productive power of colonial epistemic practices and of archiving remains active, it is also limited in specific ways with regard to acoustic collections. Until very recently, recordings were sequestered in silos of disciplines (Hamilton and Leibhammer 2014). Despite this, commentary, requests, pleas, and critique have survived in the archives as acoustically documented, performative texts. As inheritors of the archive, we can listen to these from the seams of intelligibility, that is, sometimes unable to process their content, and often ignorant of historical discourses which resonate through them, yet hopefully aware of the coloniality of the time of their recording.

Taken together, the recordings of the Lautarchiv present an enormous oratory: a massive, multivocal accumulation of songs, stories, narratives, example sentences, riddles, accounts, poems, staged conversations, and requests. Together, they form an acoustic monument to World War I, of which I can present only a section.

Do You Hear Me?

In the recording situation, Abdoulaye Niang assumed a position as speaker. The appellative power of his voice, strangled in the written files, resurfaces in the acoustic recordings when listened to by someone who is able to understand. As with many historical voice recordings that stem from projects of colonial knowledge production, the translator, in this case Serigne Matar Niang, may have been the first person who listened to and understood the speaker's request, commentary, or story in more than a century.

1.4 Abdoulaye Niang, photographed in the POW camp at Wünsdorf, Germany. Department of Evolutionary Anthropology, University of Vienna.

What does it mean to listen again, to listen closely to those recordings? Close listening is an attempt to listen to all acoustic signals. It also means to listen with the aim of understanding what is said or sung, which may entail translation and attention to the repercussions that may come from elsewhere (Hoffmann 2021). Close listening starts with foregrounding, taking acoustic signals seriously as carriers of what *can* be known by ear. This may include background noise or the sometimes faintly audible presence of speakers who do not appear in the documentation. Close listening means listening to all acoustic signals, to everything audible on a recording, and it means sometimes also noting what is not audible. It entails listening to the contradictions that emerge when what the written file indicates does not correspond with what the speaker said, and vice versa. It means listening, at best collectively, to whispers in the background, to awkward pauses, laughs, coughs, and to acoustic signatures forced onto the record by a speaker who shouted his or her name into the funnel after the song or narrative was completed (Hilden 2015; Lange 2020). Close listening often shows that the written label does not cover what is actually contained in the historical recording. Most records carry not merely "the story" or the desired language recording that the label indicates but also an announcement by the recordist and perhaps an interruption by someone else, background noises, and the sound of the recording device (see Hoffmann 2018, 2023). In other words, most recordings are composite sound objects that may "speak" in several ways. This is why I insist on writing "recordings *with* prisoners of war" instead of "recordings *of* prisoners of war." Once we listen to everything, there may be a polyphonic quality to many of the recordings, which goes beyond the polyphony that comes with particular genres (Hoffmann 2015b). Close listening may entail translation and interpretation of texts, songs, and genres. It also means listening to the recorded human voice as a socio-acoustic phenomenon, to its performative traits and abilities, and thus to what voice transmits beyond transcription. Translation, in combination with knowledge about the situation of the recording act, may allow listeners to hear moments of attempted interpellation, or aspects of failed communication, as it were, that have ended up encapsulated in a linguistic recording like butterflies pinned behind glass. With regard to the recordings of the Lautarchiv, close listening often leads to a search, to following the trace of a voice recording—sometimes out of the archive that houses the recording, perhaps to another archive, or to the clues that unwritten archives of orature may provide. With regard to most of the collections of colonial linguistics, close listening means listening together with speakers of languages spoken, with experts on particular genres, repertoires, and histories. In the case of the recordings of the Lautarchiv this means that what has been recorded by German linguists, together with hundreds of speakers of many different languages, can be researched only with the support of many translators and interpreters. In this case, close listening entails listening collectively and leaving the archive.

    In the case of the recordings with Abdoulaye Niang, his recorded voice becomes the connective tissue between archives. Niang's archival echo, the acoustic trace of his presence in the POW camp at Wünsdorf, took the lead in my search for traces of his presence in Germany and Romania. While his songs carry resonances from outside the camps, his repeated request not to be deported firmly situates him inside the camp in Wünsdorf. His recordings were listed as *Erzählungen* (narratives). His echo also leads to the archival traces of autopsy, the tenacious, gory leftovers of measuring, examining, and making plaster casts of imprisoned men. The last part of this chapter follows the recorded voice of Abdoulaye Niang from the Lautarchiv to other collections elsewhere. In so doing, it takes seriously what his recorded voice and words are still able to tell after a century. As in every chapter of this book, this also entails looking at depictions in adjacent visual collections and reading through a paper trail of catalogs, correspondence, and publications that speak of some of the ideas of the recordists and of the practices of recording.

Whether and when speakers were able to decide what they expressed in the recording sessions of the Kommission, and how things were said or sung, was related to the language spoken, to the requests of the recordists, and the interests of the person who was recorded on that particular day. Sometimes what could be said, or what was said, corresponded to the presence of others—to bystanders, an audience, or a group of speakers—during the recording session. Whereas with British soldiers, for instance, it was easy for the linguists to request recitals of specific, comparable texts, insurmountable language barriers and the differing cultural backgrounds of the African soldiers whose languages were of interest for the survey often did not allow the linguists to communicate such requests. The result was a dubious "liberty to speak" for some speakers—an opportunity to convey or to try to communicate things. Often it remains unclear whom the speaker addressed, or who the intended or anticipated audience for specific songs or messages was (Hoffmann 2009b). Where did speakers expect those records to go? Did prisoners sing or speak to each other in recording sessions during which several men were present?

Sometimes the written files provide useful clues to these questions. For example, about the recording session of December 8, 1917, consecutive files impart that three other men from West Africa were recorded along with Abdoulaye Niang (PK 1114 and PK 1115): Jámafáda (PK 1116), Adjovu (PK 1117), and Buru (PK 1119/1118). For all the recordings with these three men, very few details were given in their respective personal files, which Doegen had filled in. All three men have no surnames in these files, and the spellings of their first names seem doubtful. Yet, having no alternative options, I have to refer to the speakers by the names that were given to them at the moment their recorded voices were archived.

Jámafáda, the files say, came from Fada N'Gourma (now a town in Burkina Faso). He was twenty-one years old and had lived in France since 1915. Adjovu is listed as twenty-two years old, from Gbaye (in what was then Dahomey), and had served in the French army since 1915. For Buru, who is listed by Doegen as a "Baule Mann," and who seems to appear in Josef Weninger and Rudolf Pöch's collection in Vienna as "Goli Bru," no further information is provided in the files of the Lautarchiv.[36] All four men were recorded between 11:00 and 13:20 on December 8, 1917. Adjovu, the files say, improvised a song about the railway (the language is unclear). Jámafáda, like Abdoulaye after him, spoke of the war in Mòoré (specified by Doegen as "Mossi"; see Fragment II). Here, as in a song sung by Abdoulaye Niang and in the songs sung by Josef Ntwanumbi ("Twanumbee") (Hoffmann and Mnyaka 2015), the trains mentioned may have referred to the moment of departure from home, or they may have signaled the hope of return. Both Jámafáda and Abdoulaye spoke of their experiences of being sent to the war in what may have been a conversation between the men that was not registered in writing.

In the few cases where a linguist such as Carl Meinhof showed interest in the content of the songs or spoken texts, he was confronted with genres, which were opaque to the researcher, even if he understood some of the language spoken. On a page attached to the personal file 1490 (dated October 25, 1918, for a speaker whose name is listed as Bilali Nayga), Meinhof wrote:

> Die Übersetzung dieser Lieder macht unüberwindliche Schwierigkeiten, da sie nach Angaben des Gewährsmannes in einer besonders poetischer Sprache gesungen sind, die von gewöhnlicher Sprache durchaus abweicht.
>
> The translation of these songs is hindered by insurmountable difficulties, because they, according to the informant, are sung in a particularly poetic language that differs from ordinary language.

Yet, there were situations in which speakers articulated requests that they must have wanted to be understood. Thus, next to the uncanny choir of soldiers of the British army composed of ninety-four separate spoken versions of the biblical story of the prodigal son in dozens of English dialects, among a bewildering mix of counting, songs, and narratives, one finds sentences in the collection of the Lautarchiv that, once heard and translated, still reverberate with the appellative quality of the human voice. While the story of the prodigal son was not recorded in any African language, formulaic word lists designed for linguistic research and examples of pronunciation do exist in African language recordings, too, albeit in a smaller number. Instances of relatively free speech appear regularly. Some prisoners, like Abdoulaye Niang, apparently used the opportunity of the recording session to articulate critiques or requests.

For those who do not understand Wolof, Mòoré, or Bamanakan, the meaning of these texts emerges with their translations. Once translated, their semantic content, and in some cases, even their performative mode of speaking, resurfaces. Clearly, many recordings kept at the Lautarchiv were simply never the examples of languages they were archived to be. Instead, recordings, of "free speech" in particular, were always both examples of/ for languages *and* narratives, songs, laments, requests, and accounts that were archived mostly in complete ignorance of their semantic content and context.

Without being asked to do so, some prisoners spoke their minds, or articulated what they wanted to convey during the recording sessions held in the camps or at the studios of the Odeon recording company in Berlin Weißensee. Britta Lange writes of the prisoner Sundar Singh's request for a blanket to cover the Holy Book, expressed in a recording session held on January 5, 1917, filed as PK 676 (Lange 2011b, 122–28). While Sundar Singh was understood by the recordist Helmut Glasenapp, who noted Sundar Singh's request in the written document attached to the recording, Abdoulaye Niang's urgent plea fell silent in the archive. Personal file number PK 1114/2 lists *eine Erzählung* (a narrative or story) spoken by Abdoulaye Niang, recorded at the Odeon studios at 11:30 on December 8, 1917. Yet "a story" was not what Serigne Matar Niang heard when he listened to the recording (Berlin Lautarchiv, PK 1114/2) a century later. His translation brings out an air of urgency:

> I am truly worried as where we are right now is freezing and very uncomfortable. We were on parade this morning and the lieutenant summoned the prisoners and searched our belongings with the intention of finding money. A guy called Alexandre had 16 francs in his suitcase (*denga?*) *is that understood?*
>
> Right now, the cold is terrible, and as we are headed for [Romania], we are uncertain what we will find there. We prefer to remain where we are right now. Irrespective of our next destination, at least we are familiar with our surroundings and we are coping with the climate.
>
> May this war end, so that we can return to our parents, resume our duties and treasure them. This place is freezing; we don't know this kind of weather; we are also not used to it.

In the following two recordings the urgency intensifies: on recording PK 1114/3 Abdoulaye speaks again of the cold and the fact that the men who came from warm countries not used to it. In recording PK 1114/4 he speaks very fast:

> Warguena wakh thi kilifa gui, souniou kilifagou reuy gui! Bouniouye dieul nitgni, diyobu rebelyi, bougniou yobou Nioun dem roumanie neexuniou. Nit daal boneckee prisonier (deungako?) Te founeck niou yobulafa (deungako?), lolou dey neexul.

Tog filedaal moniou gueuneul. Seddbi barina, mettina lol Bonekkee prisonnier tam, founeka niouyobulafa, donul lila geuneul.

You must talk to the boss, our big boss! When they do select troops, as rebels, [ask] that we be exempted. We fear Romania. Once you're prisoner, did you hear? [*deungako?*] We prefer not to be moved from place to place. We want to stay here. The cold is extreme and very uncomfortable. No prisoner wants to be moved around . . . did you hear? [*deungako?*]

May this war end, so that we can return to our mothers' and fathers' houses, resume our duties and treasure them. This place is freezing; we don't know this kind of weather; we are also not used to it.[37]

Three times Abdoulaye Niang interrupted the flow of his words with the question, "Did you hear me?" or "Is that understood?" It is significant to note that in Wolof "hearing" (rather than "seeing") refers to understanding and knowing.[38] The files at the Lautarchiv do not convey comprehension.

The documentation for the recordings PK 1114/1–4 imports the obligatory details: Doegen noted that Abdoulaye Niang was born on the island of Gore (Gorée) and then moved to Dakar, where he went to Arabic and French schools. Doegen further noted that the speaker was thirty-nine years old, spoke "Wolof, Bambara, Susu, Ful, Französisch, English," and was able to read and write French and Arabic. As for the speaker's profession, Doegen wrote *Maurer* (mason), and as a religious denomination *mohammedanisch* (Muslim). Further he listed that Niang belonged to the *Volksstamm* (tribe) of the Wolof. The document also holds Abdoulaye Niang's handwriting: he wrote his name in Arabic and, next to this, in Latin script on the sheet (see figure 1.2). The file is one of the few *Personalbögen* I have seen that bears the signature of a prisoner. Read today, these signatures look like small intrusions, etched into the trail of documentation so as to claim subjectivity in the apparatus of knowledge production. Or was Abdoulaye Niang trying to prevent the misspelling of his name?

How to Formulate a Plea

Abdoulaye Niang's repeated, pressing request not to be deported to Romania does not surface in documentation of the Lautarchiv in any way. The appellative quality of his voice met no response; he was not heard by those whom he asked for help. Interpellation in its most basic sense (rather than in the Althusserian sense)—as an address that requests a response or reaction—was clearly suspended. The last question on the preprinted *Personalbogen*, as with all recordings of the Lautarchiv, required a description of the character of the voice (*Beschaffenheit der Stimme*). This was filled in by Doegen even though he could not understand a word Niang had said. Doegen describes Niang's voice as *kräftige helle Stimme mit guter Konsonanz* (strong, bright, melodic voice). His remark at once reduces Niang's voice to its sonic materiality and severs performativity from speech. Doegen's description of the prosodic features of Niang's voice does not connect with the content of his words, nor does it note the insistence with which Niang's plea was articulated. It thus isolates sound from semantics and atomizes the very composition of human speech. Disjoined from its performative plasticity, Abdoulaye Niang's voice becomes, paradoxically, both individualized and reduced to its materiality. Yet it is not individualized in the sense in which every human voice is unique—this cannot be done with three adjectives. Instead, his voice is described as a specific sound in a way that can neither account for the complexity of the acoustic fingerprint that makes for the personal voice nor admit to the social quality of voice as a cultural performance. The possibility that a language, a specific register of speaking, or a particular intention might request a specific intonation does not appear to Doegen. This, his statement explains to the reader, *is* Abdoulaye Niang's voice, no matter what he says or wants to express. It is bright, strong, and melodic.

This was not the last instance in which a researcher described Abdoulaye Niang's voice as an asemantic acoustic phenomenon. Two decades later, in 1939, the recordings with Abdoulaye Niang's song of the French recruitment campaigns in Dakar ("these are sent to the abattoir"), which Doegen cataloged, trivially, as "songs with clapping," became samples in a professorial dissertation titled "Klangstile als Rassenmerkmale" (Styles of sound as characteristics of race) by the German musicologist Fritz Bose.[39] Bose's research built on the recordings with POWs held at the Lautarchiv. The research, again, completely ignored the lyrics of the songs, which spoke of being sent to European battlefields. Bose's (1943/44) findings appeared in the *Zeitschrift für Rassenkunde und der vergleichenden Forschung am Menschen* (Journal for racial science and comparative research on peoples). The journal was published by the anthropologist Egon von Eickstedt, who had conducted racial research on Indian POWs during World War I.[40] For Bose's research, both Abdoulaye Niang's and Mamadou Samba Diallo's voice recordings were listened to by a group of test people (Bose does not specify who, nor by how many people) as examples of the sound of voice and musical expression in relation to race. It is safe to assume that the listeners did not understand the languages they heard. Based on these listening exercises, particular characteristics of voice were then attributed to the recorded voices of people who were described as either "Indianer silvider Rassen," "Neger sudanider und palänegrider Rasse," or "Europäer der nordischen Rasse."[41] Bose concedes that the sound and performative styles of both the singing and the speaking voices are related not only to specific languages but also to the gender (214) and class (213) of the speakers. Despite this differentiation, his table on the various acoustic styles of singing (*Klangstile*) conveys a clear racial distinction in the categories "Stimmklang" and "Vortragsstil" (sound of voice and style of performance). Unsurprisingly he describes the "Nordic" voices as "bright, clear, smooth, shining"; their style he defines as "cool, controlled, rather soft" and "constant." The voices of the African men he characterizes as "squawking, trumpet-like, husky, and velvety," and their style of performance is "loud, impulsive," and so on (221). Jennifer Lynn Stoever has described similar contrasting qualities of voice, based on published historical reviews of the vocal qualities of a white and a black opera singer in the antebellum US North—Jennifer Lind and Elizabeth Taylor Greenfield, respectively—as a "synesthetic crossover between visual referents to race and the sonic color line" (Stoever 2016, 103).[42] While Bose's results are explicitly racist (and he does not have to see the performers to make his racist assertions), one would have to systematically list and analyze all the characteristics Doegen attributes to the voices of several hundred singers and speakers to formulate a clearer picture of how Doegen's description of voices relates to his ideas on race. What his notes on voice do show, without doubt, is that he was neither aware of nor interested in the plasticity of vocalization, which changes with specific, distinguishable genres of speech and song. Here, as elsewhere, the speaking voice of racialized subjects became an asemantic *phone*, severed from any meaning that the speaker or singer tried to communicate.

Yet the acoustic and written documentation of Abdoulaye Niang's attempt to communicate a request allows more clarity than is possible with recordings of most other speakers. Niang appears to have assumed a position as spokesperson in the moment of recording. Niang's request not to be deported makes sense only if it was addressed to someone outside of the group of West African prisoners with whom he was recorded. His appeal, "You must talk to the boss, our big boss! When they do select troops, as rebels, [ask] that we be exempted!" clearly was an attempt to convince Doegen and Meinhof to intervene on the prisoners' behalf to prevent their imminent deportation.

To analyze the efficacy of prosodic features, melody, and wording in the recording in which Abdoulaye Niang tries to appeal to Doegen and Meinhof to prevent his and his fellow prisoners' deportation, I turn to the work of the linguist Judith Irvine. Based on her research in the 1970s in a town in Senegal about 100 kilometers from Dakar, she writes that specific modalities of speaking Wolof respond to and transmit social status and are related to the situatedness of a person in a stratified society (Irvine 1990, 128). These variations of speaking, pronouncing, and intonating Wolof—in terms of semantic and lexico-grammatical expression, tempo, melody, and pitch—she describes as "registers of language." What Irvine understands and analyzes as registers of language differs from accents insofar as the former are not determined by geographic distribution but instead signify class- and status-related specificities of speech. Irvine stresses that the communication of affect is generally not individual, nor merely "impulsive," but cultured; it is always shaped by and responds to a community of speakers (and perhaps extends even beyond this community). Expressions of affect are structured within a social and cultural framework of speech and behavior that is, in this case, strongly related to a socially stratified society (see also Abu-Lughod and Lutz 1990). Here, speech registers are determined by caste and social activity. They differ in terms of grammar, prosody, the choice of words, the tempo of speaking, as well as in the tone and volume—in short, with respect to every feature of linguistic and vocal expression. These variations were named and specified by the Senegalese speakers themselves, who differentiated two distinct registers of speech as *waxu gewel* and *waxu géér*. According to Irvine (1990, 131), one of these (*waxu géér*), a laconic way of speaking, was conventionally associated with high-ranking nobility, because these persons were assumed to be its typical users; the other style (*waxu gewel*), hyperbolic and high in affectivity, was conventionally associated with lower echelons, particularly the griots (praise singers, speech makers, and bards). The possibility of code or register switching is also important for understanding Abdoulaye Niang's recorded speech act. Irvine (1990, 136) writes that every Wolof speaker could speak in the register of *waxu gewel*, even if they did not identify as *gewel*. This, for instance, was the case when a person of higher rank used the *waxu gewel* register of speech to ask for a favor from a person of equal or higher status.

My summary of strategies of speech may sound somewhat static, and certainly spoken Wolof has undergone alteration and modification since the time of Irvine's research in the 1970s and, even more so, in the hundred years that have passed since Niang was recorded. Our position in the present, when listening to historical sound recordings, calls for caution in the assessments we can make. Yet Serigne Matar Niang, who translated and interpreted the recording, was sure he heard Abdoulaye Niang using the registers of *waxu gewel*, even though he did not belong to the griot caste. Despite the audible use of intonation and a specific way of speaking, Serigne Matar Niang contended that it is unlikely that Abdoulaye Niang belonged to the caste of griots because of Niang's education level and profession, which shine through in the files of the Lautarchiv and in the notes of Josef Weninger.

Clearly, Abdoulaye Niang found himself in a situation in which he wanted, or perhaps was chosen by other prisoners, to articulate a plea, the reception of which was determined entirely by his ability to attract the attention of the linguists. Did he expect Meinhof to understand him? Niang's articulated emphasis on being a prisoner—*once you are a prisoner, did you hear?*—suggests that he positioned himself with a clear awareness of his subaltern position in the situation. This would, in the practices of his language that Irvine describes, have been a moment of (attempted) communication calling for *waxu gewel*. In the acoustic file, this appears as fast, loud speaking, punctuated with insertions—*You hear me?*—that call to attention, together with the choice of expressive adjectives that seek to transmit a lively depiction of an extreme situation (suffering the cold). Thus, while Niang assumed a speaking position from which he addressed the recordists, he bowed performatively, using the registers of speech he had at his disposal in the hope of being heard. Additionally, the iterated usage of the term "we" suggests that Niang may have taken on the role of spokesperson, or perhaps he had been chosen to mediate because he was articulate and educated, or because he was senior to the other soldiers who were recorded on that day. This position as an intermediary who formulates a request on behalf of others, according to Irvine, would have called for the register of *waxu gewel*. What may have escaped Abdoulaye Niang's attention was the structural suspension of communication in the very moment of recording speech, which led to the failure of his attempts to effect a response with his repeated, genre-specific appeal for help. The recording freezes this moment of failed communication. In the archive it becomes a linguistic specimen that only now thaws, becoming performative speech again in the moment of translation, which, for this plea, comes one hundred years too late.

The correspondence between Carl Meinhof and Felix von Luschan complicates the story of an unsuccessful, one-sided attempt to communicate, hindered by a language barrier and the figurative wax in the ears of the linguists. Interestingly, although this surfaces neither in the written files of the Lautarchiv nor in the work of Fritz Bose twenty years later, some things that Abdoulaye Niang said *were* actually understood. A letter by Meinhof to von Luschan, written five days after the recording with Abdoulaye Niang took place, indicates that Meinhof must have understood at least parts of what the speaker said. Without naming Niang, he writes:

> Unter den Wolof ist ein gescheuter [*sic*] Maurer, der außer mehreren afrikanischen Sprachen auch französisch und englisch spricht und zwei Lieder improvisierte, in denen er etwa folgendes aufführte: "Wie ein Straßenjunge jeden Zigarettenstummel aufliest, den er auf der Straße findet, so haben die Franzosen den letzten Mann aus Westafrika aufgelesen und in den Krieg geschickt. Es sind nur noch Frauen und Kinder da. Das ist jetzt eine Zeit des Weinens. In den Städten sind die Männer weg und nur noch garnisonspflichtige Gesellen sind zurückgeblieben, die spielen jetzt den großen Herren, aber wenn ich zurückkomme, und solch Kerl kommt in mein Haus und will große Worte reden vom Krieg, dem fahr ich an den Hals!"[43]

> Among the Wolof there is a smart mason who, apart from several African languages, also speaks French and English. He improvised two songs in which he said something like the following: "Like a street kid picking up every cigarette stub from the street, the French have picked up the last man from West Africa and sent him off to war. Only women and children are left. It is a time of crying. In the cities the men are all gone and only enlisted men are left and play big men. If I come back and such a guy comes to my house, bragging about the war, I will strangle him."

This is not exactly what was recorded by Abdoulaye Niang. In particular, the moment of picking up cigarette butts and being picked up by the army seems to be twisted in Meinhof's summary. However, while Meinhof roughly recapitulated the content of what Niang sang, he left out Niang's urgent plea. There are four possible explanations I can think of: First, perhaps Niang presented a summary of what he had said in French but left out his request not to be deported, possibly because he could not muster an appropriate way or the right words to communicate this to Meinhof. Second, perhaps Niang summed up everything he spoke into the gramophone funnel in French, but Meinhof did not wish or care to communicate Niang's plea to von Luschan. The third possibility is that Meinhof understood enough Wolof to sum up most, but not all, of what Niang presented in Wolof. Or, last, Meinhof understood everything Abdoulaye Niang said in Wolof, including the prisoner's plea, and did not care enough to respond or to recapitulate the request in his letter. The archive does not allow me to determine which explanation is correct. Yet the presence of acoustic files and Meinhof's letters, in combination, emphasize again the difference between reading about something that was said and hearing it said, which in this case also pertains to the discrepancies in what was said for the record, between the researchers, and between the researchers and the prisoners.

It is not possible to know exactly for whom Abdoulaye Niang spoke. Was the prospect of deportation to Romania distressing to all the prisoners in Wünsdorf, or was it a concern only for the four men who were recorded together on that day in December 1917? All four men were interned in Wünsdorf, in the Halbmondlager (Half Moon Camp), where the living conditions were relatively good. Wünsdorf was a camp established for Muslim POWs in Germany, with the strategic aim of recruiting them for the Central Powers, of which the Ottoman Empire was an ally.[44] In Wünsdorf a mosque had been built for the POWs, and prisoners were able to live according to their religious conventions with regard to food, prayer, and hygiene. Perhaps the men feared they would have less favorable conditions of internment elsewhere?

Like Niang, Jámafáda, who was recorded fifteen minutes later, spoke of the war and of his fear of never seeing his family again. Again, Meinhof gave a short recapitulation of the Mòoré text presented by Jámafáda in his correspondence with von Luschan. As with Niang, it remains unclear whether Meinhof understood the language, or whether the speaker summarized in French what he intended to speak into the gramophone funnel. Yet with regard to Jámafáda, Meinhof's attention focused on the prisoner's appearance, which he described as "anthropologically interesting" (*anthropologisch interessant*). In his letter to von Luschan, Meinhof reported on Jámafáda's facial features, his complexion, and the scarification on his face and told von Luschan where to find the prisoner.[45] Doegen would later publish photos of Abdoulaye Niang, Jámafáda, and Buru, all of them wearing army coats. Perhaps the photographs were taken on the same winter day in December 1917 when the men were recorded. Doegen does not date the images, nor does he identify the men by name in his books. In Doegen's publication, the three prisoners become "a dark colored Woloff [sic]," "a Mossi with peppercorn hair and tribal scarifications," and "a Baule negro with artificially filed front teeth" (Doegen 1925, 32). In Doegen's visual representation, the three men are racialized, depersonalized, and unnamed.

The anthropometric files in the archive of the Department of Evolutionary Anthropology in Vienna, Josef Weninger's publication, Meinhof's letters, and the photographs published by Doegen can be linked via (misspelled) names and acoustic recordings. The connecting details, which point to a network of archives, speak to what I understand as a *Verwertungsmaschine* that operated in Germany in the prewar and war years. I understand *Verwertungsmaschine* as the "apparatus of utilization," or exploitation, that objectified people of color as exotic and racialized bodies. Visual documentation but also auscultation — a listening to language through the body of an informant, who was prompted to speak into the recording device yet who had no part in the interpretation of "language or voice material" — created a plethora of contradictory and sometimes connectible documents. The *Verwertungsmaschine* was fueled by colonial knowledge production but also by aesthetic interests and a hunger for the "exotic" that was shared by linguists, anthropologists, phonologists, musicologists, directors of *Völkerschauen* and cabarets, by artists, photographers, the general public, and those who were producing propaganda against the colonial soldiers of the Triple Entente in the German *Kaiserreich*. And, while the *Verwertungsmaschine* was already in operation before the war, there was no escape from it for the captured soldiers of World War I.

The Remainders of Autopsia

As one element of his intricate archival trace, Abdoulaye Niang's urgent plea not to be deported became an unintended connective fragment that led me to retrace a network of researchers and archives linked via their research with POWs in Germany, Austria, and Romania. As mentioned above, in 1917, Felix von Luschan, himself Austrian and the leading anthropologist in the KPPK, had invited his Austrian colleague and former student Rudolf Pöch to examine the African soldiers in Wünsdorf. Rudolf Pöch and his assistant, Josef Weninger, who together conducted their own research on POWs in Austrian internment camps, came to see the men who were interned in Germany. During their visit to Wünsdorf in September 1917, the Austrian researchers began their physical examination of African prisoners. However, most of the African and Indian soldiers who were of interest to the Austrian anthropologists were subjected to the procedures of anthropometric research in Romania in September and October 1918. This makes clear that by the time Abdoulaye Niang articulated his request not to be deported, two thousand African and Indian prisoners had already been sent to Romania. They had been deported in March of 1918 (Kuba 2015, 108). Apparently Niang (and other POWs in Wünsdorf?) saw the deportation as a punishment and not, as official versions would have it, as a measure to prevent more damage to the health of African and Indian prisoners who were not used to the cold climate in Germany.⁴⁶ Pleading to be exempted from deportation, Niang mentioned the biting cold of the Berlin winter that the Auswärtige Amt (Foreign Office) brought up as the definitive reason for their deportation to Romania. Twice Niang affirmed that the cold was indeed harsh, and that the men were not used to this climate, only to subsequently emphasize that they would nevertheless prefer to remain in Wünsdorf. His recording seems to respond to, and resonate with, a discussion around the deportation of prisoners. It addresses the official reasoning that was given for their deportation and presents an acoustic glimpse of what may have been speculation circulating in the camps that have so far, to my knowledge, not appeared in writing.

Abdoulaye Niang's recorded plea not to be deported points to Josef Weninger's publication of 1927, *Eine morphologisch-anthropologische Studie: Durchgeführt an 100 westafrikanischen Negern, als Beitrag zur Anthropologie von Afrika* (A morphologic-anthropological study: Conducted on 100 West African negroes as a contribution to the anthropology of Africa). The publication is based on the research of Rudolf Pöch and his assistant Josef Weninger in POW camps in Germany and Romania, the results of which were published after Pöch had died in 1921. Weninger's publication shows anthropometric mug shots, describes bodies, and presents racializing characterizations of African men, whom he subdivides into ethnic groups. Abdoulaye Niang features as number 57/3964.

Between 1915 and 1918, seven thousand POWs were examined in Austrian, German, and Romanian camps. Pöch and Weninger produced five thousand anthropometric photographs, around two hundred plaster casts, ethnographic films, phonographic recordings, and a collection of hair samples (Berner 2004, 607; Lange 2013).

Amid the debris of imperial knowledge production kept in the archive of the Department of Evolutionary Anthropology in Vienna, one finds grotesquely atomized representations of bodies, irreversibly depersonalized and effaced by practices of autopsy, visual representations on which racialization was based. In Vienna, there is no text in relation to the men who were subjected to these examinations in Romania and at Wünsdorf. Only one comment in Weninger's writing allows the prisoners' resistance against this violent apparatus to surface. It is apparent that some of the African prisoners were photographed with a white cloth covering their shoulders and chests. Their refusal to have mug shots taken with bared upper bodies filters through in Josef Weninger's (1927, 3) expression of surprise vis-à-vis their unexpected resistance in this moment.

1.5 Austrian anthropologist Rudolf Pöch's list of measurements of Abdoulaye Niang (detail). Department of Evolutionary Anthropology, University of Vienna.

In the archive of the Department of Evolutionary Anthropology in Vienna, the dehumanization of the prisoners is complete, irreversible, and irreparable (Scott 2018). Of Buru, now named Goli Bru, who was present when Abdoulaye Niang tried to petition against their deportation, one finds only measurements.[47] Again, he is registered as speaking no language other than "Baule" (Baoulé). Together with Abdoulaye Niang, he had previously been examined in Wünsdorf. Samba Diallo, now registered as Samba Dschalo, who sang of the war as the "Catcher of the living," was caught in the grid of examination too. The files in this Viennese archive provide the date and place of Niang's examination (see figure 1.5). In his anthropometric descriptions, Pöch notes that Niang has an injury on the left foot, that he has traveled a lot and speaks several languages. Also in this documentation, Abdoulaye Niang becomes a trader, while for Meinhof he was a "clever mason."

In his book, Weninger claims that Niang was married and had one child (1927: 157). *Konskriptionslisten* (conscription lists; see figure 1.6) give an overview of the examinations and specify the practices of visual representation to which the prisoners were subjected. Of the photographs taken of Abdoulaye Niang, only one has his name written on the back. On the others, Pöch's handwriting registers him as "Wolof" or "hamitic negro." The section related to bodily features lists dozens of bodily measurements as well as the color of his skin and hair according to one of the usual scales of anthropometry.

In the registers of the anthropologists, I find traces of the presence of Paul Panda Farnana (1888–1930), the later prominent Congolese-Belgian intellectual and activist, who endured the entire apparatus of examination and anthropometric depiction in the camp at Turnu Măgurele in Romania.[48] Paul Panda Far-

Anthropo-logische No	Kriegs-gefangenen No	Name	Volksstam
3951	11/oa	Bel arbi Abdollader	Araber
3963	5/104	Mohammed Ben Redjeb	Araber
3964	4/296	Abdoulaye Niang	Wolof
3965	36515	Amady Amadou	Damga

1.6 Rudolf Pöch and Josef Weninger's *Konskriptionsliste* with data about Abdoulaye Niang (detail). Department of Evolutionary Anthropology, University of Vienna.

nana had volunteered for the Belgian army and was taken prisoner in Namur, southeast of Brussels, together with Albert Kudjabo. He was deported before he could be acoustically recorded, which Meinhof stated with regret in his report to the KPPK in February 1919. In the documentation of racial research in the archive in Vienna, Farnana, like everyone else who had undergone the examination, surfaces as a virtually dissected body, stripped of personality and reduced to an example of an imaginary racial entity. In his case, a plaster cast of his head was also produced, which was later used as teaching material at the Institute of Ethnology and Anthropology in Vienna. The racializing visual practices of anthropometry add nothing to the knowledge of Farnana as a historical figure, apart from clues about the humiliation he experienced. From the camps, Farnana appears to have established contact with the Senegalese member of French Parliament, Blaise Diagne. Farnana began his correspondence with Diagne following conversations with other prisoners about recruitment campaigns and the conditions of imprisonment. Whether Paul Panda Farnana and Abdoulaye Niang met in the camp in Romania is not known.

The Department for Evolutionary Anthropology in Vienna also holds a stereo photograph of Abdoulaye Niang, on which bacterial damage has left a golden halo around his head and face (see plate 1). Seen through the viewing device, the stereo photograph produces a three-dimensional presence, a presence that is both uncanny and almost real.[49] Pöch's stereo photographs

1.7 Death certificate of Abdoulaye Niang, issued by the French army, July 17, 1919.

attest to his eagerness to use all available media to depict the people who were subjected to his research. The stereo photographs also supported Weninger's idea of developing a "morphologischer Blick," a gaze that documents the shape of bones under the skin so as to create typologies and categories of race (Weninger 1927, 16). Seen as a personal trace of Abdoulaye Niang, the stereo photograph's uncanny clarity transmits an air of melancholia, and perhaps the signs of the illness that would not allow him to return to Dakar.

The question of whether Abdoulaye Niang was deported to Romania is not answered in this archive. Yet in the collection of the Frobenius Institut in Frankfurt, Niang appears in an undated group photograph that shows men digging (see detail in plate 2).[50] Frobenius, as mentioned earlier, had overseen the camp Turnu Măgurele in Romania. No further information is given on this picture. The document that designates the end of Abdoulaye Niang's journey is his death certificate, issued by a hospital in Lyon and now published online by the French ministry of defense (see figure 1.7).

After being released from Turnu Măgurele, Abdoulaye Niang died on July 17, 1919, from tuberculosis, which he had contracted during military service. His echo has the last word (Berlin Lautarchiv, PK 1115/5):

I cry and I cry and I cry, only I know why.
When the train stops at Ndakaru Dial Diop [train station in Dakar]
    you can prepare me for my burial.
That day, mother, I am too happy, you hear me?
Any girl who wants to date me, I insult her!
You must try to hear me!
All the men went to that war.
Any girl—you hear?—who wants to date me, I will insult her![51]

Fragment III

Asmani ben Ahmad: "Once upon a time"
TRANSLATED FROM KISWAHILI BY DISHON KWEYA

Once upon a time (*zamani sahale*) my father held me by the hand/took me to the Koranic school. Then he caused trouble (abomination?), which brought/caused trouble that upset the entire nation. Then he was arrested and fined. Then I left. I left (with a friend?) and we went to Mayotte. I stayed in Mayotte and then I came back. When I came back, my father had caused more upheaval in the kingdom. They wanted to arrest him, so he left and went away for good. He never came back. I stayed, me and my teacher, then I left alone and went to Mayotte. I stayed in Mayotte and then I left and went to Mwali. I lived in Mwali and then my boss (?) sent me to Buhini. I stayed in Buhini for a period of three months, I was employed as an askari. I worked as askari for a period of three months. After I was paid, I left and went to Naindjirani, for a period of seventy days. Then I lived in Naindjirani for two years. Then I left Naindjirani; I embarked on a Dhau that went to the military post at Morondava. I followed the upper road, then I met one Sikh woman. Her father is an Arab. She is looking at us nicely. She has looked at us nicely. She has cooked nice food for us, with milk. Then we went . . . then I went back on the road, I went to Morondani.

(Berlin Lautarchiv, PK 1108/1)

Asmani ben Ahmad came from Mwali, on the Comoros islands, and was born in 1892. He spoke Kiswahili, French, and Arabic. The files of the Lautarchiv state that he moved to Madagascar when he was eighteen years old. The account of his loss of home and his subsequent journey most probably is autobiographical. In his narrative of his journey, Asmani ben Ahmad seems to reminisce about what had brought him to a German camp.

Plate 3 shows prisoners from Madagascar and Comoros, together with the linguist Paul Hambruch. The photograph was probably taken in the camp at Wünsdorf. Asmani ben Ahmad is the second from left.

2

Mohamed Nur

TRACES IN ARCHIVES, LINGUISTIC TEXTS, AND MUSEUMS IN GERMANY

> [A] second topic in need for critical attention is the way an Africanist idiom is used to establish difference or, in a later period, to signal modernity... how the dialogue of the black characters is construed as an alien, estranging dialect made deliberately unintelligible by spelling contrived to defamiliarize it... how it is used to establish a cognitive world split between speech and text, to reinforce class distinction as well as to reassert privilege and power, how it serves as a marker and vehicle for illegal sexuality, fear of madness, expulsion, self-loathing. —TONI MORRISON, *Playing in the Dark: Whiteness and the Literary Imagination*

Mohamed Nur, Josef Ntwanumbi, Stephan Bischoff, Ali bin Bedja, and Toby Roberts were interned during World War I in Ruhleben, a camp for noncombatant foreign civilians, near Berlin. Nur, Ntwanumbi, bin Bedja, and Bischoff were recorded by the Königlich-Preußische Phonographische Kommission (KPPK) in April and May 1917.[1] The records (and now digital files) that resulted from these recording sessions are held in the Berlin Lautarchiv.[2] The files of the Lautarchiv do not mention any resistance to the KPPK's attempts to record languages.

2.1 (*above*) Mohamed Nur (detail of plate 5). Deutsches Historisches Museum, Berlin, Nachlass Wilhelm Doegen.

In Wallace Ellison's 1918 autobiographical novel *Escaped! Adventures in German Captivity*, an internee named Toby Roberts yells, "If you try to gramophone me, I'll kill you!" The author mentions Toby Roberts's angry outburst to demonstrate that he "became of unsound mind" (Ellison 1918, 63). Roberts appears in the chapter titled "The Darkies in Ruhleben" as one of the Africans who were interned in the so-called Engländerlager (English camp) situated on a former racecourse in Ruhleben. Up to five thousand mostly British men between the ages of seventeen and fifty-five, categorized as "foreign subjects," were interned there.[3] The historian Matthew Stibbe (2008, 98) estimates that most of the black internees in Ruhleben were seamen and fishermen. Among them were many African seamen whose ships had entered port in Hamburg when (or after) the war had begun.

What can be known of this camp and the circumstances of its internees is filtered through an archive quite different from the textual and visual archival record of the Halbmondlager (Half Moon Camp) in Wünsdorf, where Abdoulaye Niang was interned (see chapter 1). The Wünsdorf camp held prisoners from Africa, Asia, and India who were enlisted in the militaries of Germany's wartime enemies. It was where the first mosque in Germany was built, and it attracted a Berlin public eager to stare at "foreigners." Photographers and artists as well as delegations of anthropologists and linguists visited the camp in search of exotic bodies (Burkhard and Lebret 2015; Kahleyss 2015; Lange 2013; Roy, Liebau, and Ahuja 2011, among others). In Wünsdorf, artists' and anthropologists' images of the prisoners and the camp create a kaleidoscopic impression of perceived otherness—sometimes beautified, but always racializing and exoticizing. In Ruhleben, where the majority of prisoners were white and British, the inmates were subjected neither to the *Schaulust* (voyeurism) of visitors nor to the interests of anthropologists. Yet after the war, Ruhleben became a storied location: many of the British internees published autobiographic accounts of their experiences in German captivity. In addition to several novels by former internees and Matthew Stibbe's (2008) history of Ruhleben, the extensive collection of the former camp librarian, Maurice Ettinghausen, has been preserved (Ketchum 1965; Cohen 1917; E. Stibbe [1919] 1969, among others). While interned, Ettinghausen, a dealer of rare books who had lived in Munich when the war began, collected all available documentation from his time in the camp and arranged to keep the material when he was released. Today the Maurice L. Ettinghausen Collection is held at the library of Harvard University's law school. The collection includes sixty photographs taken secretly by unidentified prisoners who had been able to hold on to their cameras in the camp. The photograph

2.2 Photograph taken by a prisoner in Ruhleben camp. Possibly the exterior of barrack 13, designated for black internees. Maurice L. Ettinghausen Collection of Ruhleben Civilian Internment Camp Papers, Harvard Law School Library, Historical and Special Collections.

shown in figure 2.2 is one of these. The names of the men who sit in front of a barrack are not indicated on the photograph.

Ruhleben camp was self-governed by its British inmates, who arranged a spatially segregated scheme of accommodation, with separate barracks for Jewish as well as for black internees (M. Stibbe 2008, 59). The photograph in figure 2.2 may possibly be of internees sitting in front of barrack number 13, which was the designated camp accommodation for internees of color.

In Ellison's (1918, 58) autobiographical narrative, Toby Roberts is caricatured and presented together with other interned seamen from West Africa as "little more than savages." Ellison describes the African men he met in Ruhleben as illiterate, simpleminded, and overwhelmed by the situation of captivity. His prose presents their language as a strange, unintelligible idiom. The language Ellison chose to represent the black prisoners purposefully marks the difference between his own written text and their spoken words and points to what he conceived of as an unbridgeable distinction between himself and "them." Toby Roberts's anger thus becomes a sign of his paranoia paired with ignorance: according to Ellison, Roberts confused the (almost) assonant verbs "gramophone" and "chloroform," which supposedly demonstrates that Roberts could not make sense of either term—or of the technical

68 CHAPTER TWO

achievements that they described. Most stereotypically, Ellison represents Roberts's incomplete command of the English language as a deficit. According to Ellison, what Roberts really meant was: "If you chloroform me, I'll kill you."

Despite the book's claim to be autobiographical, one could dismiss this scene or "episode," as Ellison calls it, as poorly written, racist fiction—which it is—with no historical significance whatsoever. However, searching for traces of the African prisoners of Ruhleben, I came across a document that unsettled my earlier assumption that the text is entirely fictional. The name Toby Roberts appears on an official list of British seamen and fishermen taken prisoner in 1914–18, published in 1918.[4] The list records a man named Toby Roberts as having been interned in Ruhleben during the war. The existence in the Berlin Lautarchiv of two hundred voice recordings made with internees from Ruhleben shows that prisoners were, indeed, "gramophoned" there during World War I by the KPPK. Coming upon this information shifted the contextual frame through which I read the scene in Ellison's book.

According to the list, which was issued by the British Board of Trade and is currently held by the National Archive in Kew, a Toby Roberts was employed as a stoker on the *Dartwen*; lived in Cardiff, Wales; and was released from the Ruhleben camp in 1917. Toby Roberts's appearance as a character in the autobiographical novel and on a list of interned seamen, together with the fact that the musicologists and linguists of the KPPK had indeed made recordings in Ruhleben—albeit secretly—allows for the possibility that the anger expressed by the man called Toby Roberts in Ellison's book is real. Perhaps Roberts really said, or shouted, that he did not want to be gramophoned. If so, angry as he may have been, Toby Roberts was not paranoid or ignorant. As the Board of Trade list indicates, he was one of the merchant seamen who had been interned in the camp in Ruhleben, most likely after his ship had entered port in Hamburg. Perhaps he was asked to speak into the phonograph but refused. The files of the Lautarchiv document only successful recordings. They do not mention acts of refusal by prisoners, which would indicate the failure to acquire an acoustic document in a desired language. No recording with Toby Roberts's name is registered at the Berlin Lautarchiv. Yet the combination of his name on the list with the fact of the operation of the KPPK in Ruhleben and the possibility of his refusal allow us to question Ellison's assertion of Roberts's madness. But neither the novel, nor the list, nor the knowledge about the work of the commission allows us to know Toby Roberts as a historical figure.

Nor can Toby Roberts be identified in a photograph from the Ettinghausen Collection that shows black civilian internees in Ruhleben, though he may well be in this picture (see plate 4). According to the scheme of segregation in the camp, he would have stayed in barrack 13, the one assigned to black internees of Ruhleben, probably shown here. The group photograph swallows individuals in favor of depicting a crowd of men of color. This picture was taken by the German photographer Albert Grohs. There are captions written on the back of the photograph: one presumably written by the photographer; another probably added by the collector.[5] The men's skeptical faces belie the staging of a happy, well-fed crowd of civilian internees in a German camp, a representation that was likely intended for propaganda purposes. The depiction and staging of the group of men in the photograph differ from what one finds in photos of African POWs in German camps: these captives were not soldiers; they were not regarded as enemies. The internees of Ruhleben camp had not fought in the Great War but were swept up in the maelstrom of belligerent politics. Some were taken into custody straight off their ships, like the seaman Josef Ntwanumbi from the Eastern Cape, who was also interned in Ruhleben. "Khwel' itrain!" (Get on the train!), he sang in isiXhosa on one of his recordings. On another recording Ntwanumbi sang of hunger, isolation, fear, and uncertainty (Hoffmann and Mnyaka 2015; see also Fragment IV). Other black internees had lived in Germany for years prior to the war, like Stephan Bischoff from the Gold Coast (now Ghana), who, like Ntwanumbi, was a seaman, but whose address in the Red Cross registration of POWs is Auguststrasse 4 in Berlin (see chapter 3).[6]

Toby Roberts's angry refusal to be "gramophoned" sounds paranoid because he is depicted as being ignorant and because the KPPK's recording project had apparently remained a secret. Most probably, the author of the autobiographical narrative was not aware of the activities of the linguists and musicologists who visited the camps. The secrecy of the operation in tandem with the author's blatantly racist attitude, which dismisses the possibility that an African internee could know something of which the author was not aware, make Toby Roberts, the literary character, a madman. Toby Roberts (the man) does not speak in or through the archive. He is not identifiable in the photographs of internees of color in Ruhleben. The textual trace of what I imagine as his furious refusal to become a native informant for the KPPK is filtered through a narrative that claims him and has already turned him into a man "of unsound mind." The literary figure Toby Roberts belongs to Ellison's narrative, which is an exemplar of a larger discursive formation that Achille Mbembe has described as the egregious work of fabrication (*die ungeheure Fabulierarbeit*) that did (and continues to do) the literary legwork for the construction of races and collective bodies (for instance as "African"). This work of fabrication constantly reconstructs and verifies the borders that divide the entities created by the discourses of racial difference (Mbembe 2014, 11).

Like other colonial texts, Ellison's novel does not "unwittingly permit the recuperation of the subaltern subject" (Lalu 2009, 63). A critical reading of the figure of the black internee in the novel, along with supplementary information from other archives, destabilizes Roberts as a literary character yet cannot recuperate him as a historical person: he may not have been a madman, yet he cannot be reconstructed as a person from the archival fragments that acknowledge his presence in Ruhleben.

Would a text in Toby Roberts's name make a difference? The answer is simple: yes. But what if the text were, for instance, a biographical snippet in a grammar, already owned, as Ranajit Guha (1987) has described it, by another discourse? What if it were transcribed by another person, and thus edited? Would a voice recording be entirely different? How historical recordings speak depends on how restrictive the moment and situation of recording was, on the formal restrictions of the medium, and on the specific archival practices applied in relation to any doubtful freedom of speech in the recording situation. It also depends on how much freedom to speak the opacity of a language not understood by the recordist allowed, or how genres that allow for metaphorical speech add to the opacity of the spoken performance. In relation to the African POWs who were sucked into the *Verwertungsmaschine* (apparatus of exploitation) of imperial science, propaganda, and voyeurism, there rarely is enough material—visual, textual, acoustic—to address these questions. The situation is different, however, in the case of the extensive, complex traces of Mohamed Nur that exist in several archives in Germany and the United States.

Sounds Odd

Example sentences found in grammars may seem to be unlikely testaments to the long, unequal conversation between researchers and those who were labeled "native informants."[7] Yet linguists tapped into what they understood as local, or indigenous, knowledge to acquire information for the grammars they were compiling. They collaborated with speakers of the languages in which they were interested to formulate word lists and to learn the intonation, pronunciation, and grammatical structures of the languages. Whether these (mostly male) "informants" were the ideal "native speakers" the researchers were looking for and how such "ideal" speakers would have been defined are separate questions. The grammars and linguistic texts resulting from these encounters are part of the colonial archive, and they resonate with the attitudes of their authors. Sometimes they carry echoes of the conversations and, with these, repercussions of the power relations inherent in the epistemological practices that stood in the (more or less direct) service of colonization.

In an article on Somali vowels published in 1919 by the Seminar für Kolonialsprachen (Seminar for Colonial Languages) in Hamburg,[8] a sentence appears in which words that seem unhinged, severed from their context, speak of the opportunistic liaison between war and linguistics that constituted the basis for the collection at the Berlin Lautarchiv. Between sentences that speak of the weather, sentences that obviously serve to explain grammatical structures, one reads:

> In den zukünftigen Krieg wird niemand gehen wollen
>
> Nobody will want to join the future war

The words seem to flicker amid the tranquility of the grammatical tract. They appear in Maria von Tiling's contribution to the *Zeitschrift für Eingeborenensprachen* (Journal for native languages; von Tiling 1918/19, 144). The sentence was written or spoken (or both) by a young migrant from Berbera, the most important port city in what was then British Somaliland (from 1888) and the Gulf of Aden. His name appears in von Tiling's texts and in the files of the Lautarchiv as "Muhammed Nur" (spelled Mohamed Nur here). Before he met von Tiling, Nur had been acoustically recorded in the civilian camp Ruhleben in late April 1917. It is likely that he shared barrack 13 with Toby Roberts, Ali bin Bedja, Stephan Bischoff (see chapter 3), and Josef Ntwanumbi, the seaman from the Eastern Cape in South Africa.

Nur's sentence sounds oddly misplaced, homeless perhaps, in this grammar, because World War I, to which the sentence refers, is not mentioned in the publication. The absenting of the (then) recent war as an opportunity for research, which the KPPK had recognized in the incarceration of thousands of men from all over the world, occludes the grounds on which war and linguistic knowledge production met. The omission was not unusual; it was common practice in surveys of languages and anthropometric practices (see Irvine 2008; Hoffmann 2011a, 2020b). In this case the omission conceals the strategic opportunism of the KPPK's operation in POW camps. The propinquity of what may appear as an unlikely liaison, that of war and linguistics, demands explanation. In von Tiling's book *Somali Texte und Untersuchungen zur Somali Lautlehre* (Somali texts and research on Somali phonetics; 1925), as in the publications of the Lautabteilung (Lautbibliothek 1928, 1929; Lautbibliothek and Westermann 1936), the war is not mentioned. Yet it surfaces in the texts that were spoken by the interned men, and it is in some cases readable in voiced texts that were transcribed, translated, and turned into examples of languages by the linguists. Prisoners, or ex-prisoners like Mohamed Nur, spoke of the camps, of hunger, of injuries, and of traumatic experiences. Their expressions of suffering, hunger, or sadness did not exact any comment in the files of the Lautarchiv or in the few publications based on the recordings. The fact that certain things could be said in recordings that would have been censored in outgoing letters—for instance, the reference to hunger by Josef Ntwanumbi, who was interned in Ruhleben—makes clear that these texts were heard and read as linguistic examples with no communicative value or character (Hoffmann and Mnyaka 2015).

By 1919, when von Tiling's article appeared, Nur was already employed as a language assistant at the Hamburgisches Kolonialinstitut (Meyer-Bahlburg and Wolff 1986). At Carl Meinhof's request, Nur had been released from the Ruhleben camp by the end of 1917. On December 1, 1917, Meinhof wrote to Felix von Luschan:

> Eben habe ich . . . den Somali zu begrüßen, den ich in Ruhleben fand. Ich beschäftige ihn hier als Sprachgehilfen, er ist ein feiner Mensch.[9]

> Just now I had . . . to welcome the Somali whom I found in Ruhleben. I employ him here as a language assistant, he is a fine fellow.

At the institute in Hamburg, Nur was one of two language assistants who had been interned in Ruhleben; the other, Stephan Bischoff, was to teach Ewe. The acoustic recordings held by the Lautarchiv were produced with each of the two internees on the same day. Both were hired straight from the camp. For Nur and Bischoff, the linguists' interest in the languages they spoke meant that they could leave the camp before the war was over.

Maria von Tiling had no part in making language recordings of POWs in German camps.[10] Yet, for her, as the daughter of an East Prussian pastor and a teacher of French and German, the war had created opportunities, too. The conscription of every man at the Hamburgisches Kolonialinstitut—with the exception of the director, Carl Meinhof, who was fifty-eight years old when the war began—had caused a shortage of staff. In the absence of men, under Meinhof's mentorship, von Tiling qualified as one of the first female Africanists (*Afrikanistinnen*) in Germany. Her assistant was Mohamed Nur. Von Tiling writes:

Die kurzen Beispielsätze stammen sämtlich von einem eingeborenen Somali namens Muhammed Nur, derselbe ist ein äußerst intelligenter Mensch, der seine Sprache vollkommen beherrscht und sich der feinsten Unterschiede bewußt ist. Er ist im Jahre 1891 geboren in Berbera in British-Somaliland und gehört zum Stamme der Háber Ja'alo. Er ist Mohammedaner mit arabischer Schulbildung, etwa 6 Jahre lang, von 1903–1909 besuchte er die Schule in Berbera, danach war er zwecks weiterer Schulbildung noch fünfmal je drei Monate in Aden. Er beherrscht das Arabische in Wort und Schrift, außerdem spricht und schreibt er verhältnismäßig gut deutsch und auch englisch. Beide Sprachen hat er in Europa gelernt, wohin er 1911 gekommen ist. Seit Dezember 1917 ist er als Sprachgehilfe am Seminar für Kolonialsprachen des Hamburgischen Kolonialinstitut angestellt. (1918/19, 134)

The short example sentences all come from a native Somali named Muhammed Nur, who is a very intelligent person, in perfect command of his language, and who is aware of the slightest nuances. He was born in Berbera in British Somaliland in the year 1891 and belongs to the Háber Ja'alo tribe. He is a Muslim (*Mohammedaner*) with Arabic schooling, he went to school in Berbera for about six years, 1903–1909, and visited Aden, for further education, for five three-month periods. He speaks and writes Arabic, and speaks and writes German and also English fairly well. He learned both languages in Europe, where he came in 1911. Since December 1917 he has been employed as a language assistant at the Seminar für Kolonialsprachen of the Hamburgisches Kolonialinstitut.

Von Tiling's emphasis on Nur's intelligence and his command of his own first language, Somali, shows that African language assistants were by no means treated as equals to their German colleagues at the Hamburg Kolonialinstitut. Yet, as the files of the Kolonialinstitut indicate, Mohamed Nur was well aware of the role his expertise played in the research on Somali. On March 21, 1921, Carl Meinhof wrote to the governing authority (the Oberschulbehörde) that Mohamed Nur had requested an increase in salary after he learned that one of the language assistants earned more than he did. According to Meinhof, Nur also stated that his work and intellectual engagement allowed the institute to understand the Somali language.[11]

Von Tiling's dissertation on the Somali language is based on her asymmetrical collaboration with Nur as a *Sprachgehilfe* (language assistant).[12] All we know about their collaboration is what can be gleaned from von Tiling's publication. "We," in this case, refers to Bodhari Warsame, a Somali American based in Sweden, and me. Bodhari Warsame is working on a biography of

Mohamed Nur and has been researching the globalization of the Somali peninsula, the historical formation of the Somali diaspora, and Somali cultural heritage. We met while we were both looking for traces of Mohamed Nur in 2013. Since then, we have exchanged archival information and images. Warsame has translated and analyzed the Somali recordings of Mohamed Nur held at the Lautarchiv, while I assisted him in navigating the German files.

Nur, who spoke German, English, Somali, and Arabic, and who wrote in German, English, and Arabic, worked closely with von Tiling and taught Somali together with her until he left Germany in 1921. According to von Tiling, Nur wrote his texts first in Arabic, and then in Latin script, thereafter reading them to von Tiling, who transcribed them phonetically. At that time there was no standardized script for Somali. According to his biographical narrative, which appears in von Tiling's 1925 publication, which is based on her PhD thesis, Mohamed Nur spent most of the war in Ruhleben.[13]

Somali Texte was the second publication based on von Tiling's collaboration with Nur, following her article on vowels. It appeared when Nur had already left the country, about six years after Germany had lost the war. By then, it was apparently safe for von Tiling to mention in a publication the secret operation of the KPPK in the wartime internment camps. It was Nur himself, though, who presented an account of how he came to be interned in Ruhleben and subsequently hired by Carl Meinhof. Von Tiling acknowledged his assistance and his presence in Hamburg, yet she kept quiet about the circumstances of his employment at the institute. The publications that they crafted together appeared under her name only.

Perhaps Mohamed Nur had learned in the camp about the devastation of the Great War, for which, according to his account (written with von Tiling), he had tried to enlist. Nur had wanted to fight on the German side against the army of the United Kingdom, the colonial power that had colonized his home country. Certainly, news and rumors about the trenches and the horrors of this war had reached the camp (Ruhleben) in Berlin. Nur's example sentence, "In den zukünftigen Krieg wird niemand gehen wollen" (Nobody will want to join the future war), resonates with the terror of World War I. This was a war that was very different from the battles Nur refers to in the recordings that were produced with him as a speaker and that are held at the Berlin Lautarchiv. The appearance of this sentence in von Tiling's article of 1918/19, which does not mention Nur's experience of captivity, speaks of the practices of knowledge production, in the course of which soldiers and foreign internees became informants. It belies any claim to scientific detachment and to linguistics being a disinterested epistemological practice.

Mohamed Nur's archival trace is at once extraordinarily complex and dispersed. While many of the speakers in the Lautarchiv left little more than the archival echo of their voices and of things they said or sang for the recording, Nur's presence in Germany left a variety of images, texts, and sound files. My engagement with Nur's archival traces begins with the move of retrospectively claiming his coauthorship of the written texts that appear in *Somali Texte*. By doing this, I seek to reclassify these autobiographical snippets as belonging to another order: that of historiographic sources, rather than of grammatical examples. This move follows Ranajit Guha (1987), insofar as it attempts to wrest text and semantic meaning from grammatical examples. Yet, as discussed in chapter 1, one cannot simply reclaim such documents for history, ignoring the discourse that owned—and coproduced—these texts in the first place. My emphasis is not on clearly identifiable authorship here. Nur did not, as far as we know, sit down to write the story of his journey. His narratives were produced in response to von Tiling's requests that he provide textual material in Somali. Nur's narratives thus answered von Tiling's requests, which do not surface in her text. This makes Nur's narratives similar to the echoic voices of the recordings, in the sense that one can read in them the answers but not the questions that prompted them.

As with acoustic recordings, linguists' need for textual examples prompted narratives, which were subsequently neutralized, cast as examples of languages. Yet the trace of the productive power of epistemological practice remains within these texts. In reassembling archival snippets that correspond to one another, Nur's traces across several archives and genres, fragments of the story of his journey as well as facets of his self-representation, become readable as narratives rather than just semantically emptied language examples. As a consequence of the nature of Nur's cooperation with von Tiling, and related to the fact that she appeared in her publications as the author of what perhaps should be understood as "their texts" (which he had first written and then read to her), it is difficult to discern how much she edited these often autobiographical narratives. Nothing is known about the practical implications of their collaboration. The acoustic recordings in the Lautarchiv in which one hears Nur speak and sing are also mediated in several ways. Here, too, the requests of the researchers do not appear in the written files. All narratives and songs are abbreviated due to the length of the gramophone records or wax cylinders and by the recording situation (in the POW camp). As will be shown in chapter 3, researchers' interest in and requests for specific genres can at times be gleaned from other archival material. In the recordings from Ruhleben, as in the recordings with Abdoulaye Niang in Wünsdorf, the situation in the camps comes across in some of the comments of the prisoners that mention the biting cold. The degree of mediation and modification differs between the acoustic and the written files and also between different sound recordings. Some of the recordings are series of words that were requested by the linguists. In others, one can hear direct commentary, songs, prayers, or stories. The most significant difference between the acoustic files, the biographical fragments, and the example sentences in von Tiling's texts is the fact that in the recordings one can actually *hear* Nur chant. This means that one can reappraise his words outside of the logic of transcription within a colonial academic setting, albeit from a distance of one hundred years. And while one can hear Nur's acoustic voice on the recordings, this audible voice may not be Nur's political voice (alone). Some of his recordings include words that were his own; in other cases, he recited poetry. It is not always possible to discern if and when his acoustic voice was congruent with his political voice. When Nur cites a Somali poet in his recordings, the result is a polyphonic spoken text: one hears Nur's words interlaced with snippets of political discourses from Somalia, as interpreted by Nur.

The Apparatus of Exploitation

Apart from the written and spoken texts, there are a variety of images related to Nur that show him in different, at times incommensurable, ways. The images not only are distinguishable with regard to the technique of production (two X-ray images, two photographs, seven paintings, several drawings), but also differ immensely in terms of the scale of coloniality at play in these representations. Of the images Bodhari Warsame and I have been able to identify, only three are connected to Nur's name: a painting by Rudolf Jacob Zeller (plate 6) and two X-ray images. They appear in von Tiling's *Somali Texte* (1925, figs. 5 and 6). The other images, like the disconnected sentences in a grammar, seem unhinged from the person who was once depicted. Images of Nur form part of a larger pool of images and artworks created with black models in Europe who were rarely named or identified. Although this may have been said elsewhere and is the topic of recent exhibitions—for instance, *Posing Modernity: The Black Model from Manet and Matisse to Today/Le modèle noir, de Géricault à Matisse* (2018–19), at the Wallach Art Gallery, New York, and the Musée d'Orsay, Paris—the existence and genres of these unnamed artworks for which people of color were models must extend our definition of the colonial archive to include materials held by art museums and in private collections.[14]

While the painting shown in plate 6 clearly depicts Mohamed Nur, as he is named under the reproduction of it in von Tiling's article, we do not know much about the making of the artwork. The artist Rudolf Jacob Zeller, whose name also appears under von Tiling's reproduction, apparently did not sign the painting.[15] Dated 1921, it is today held and displayed at the Hiob Ludolf Center für Äthiopistik at the University of Hamburg, where I was able to see it in 2019. Only Maria von Tiling's 1925 publication identifies this artwork as a portrait of Nur. This means that it is only in her publication that Nur's proper name (in German spelling) appears in connection with his image. Apparently, his work as a language assistant, and perhaps his long-term collaboration with von Tiling, "qualified" him to be named in this context, whereas as an internee of color in Ruhleben, as an informant for Meinhof and Wilhelm Doegen, or as an "exotic body," he was not. In all other photographs and images we have found of him thus far, Mohamed Nur is either not named or appears under a pseudonym.[16] This painting thus forms the nexus of orientation in our search for visual traces of Mohamed Nur across a network of archives.

The other image from von Tiling's publication that is of interest for this search is an X-ray of Nur's voice box and larynx produced by Giulio Panconcelli-Calzia for the Hamburgisches Institut für Experimentelle Phonetik (Hamburg Institute for Experimental Phonetics). Visualization of language and "race," in this case, did not halt at the barrier of the skin.[17]

Using the painting in plate 6, we were able to identify Nur in several images, two of which are group photographs from the camps. In the propaganda photograph that is part of the Ettinghausen Collection (plate 4), he is located third from right in the front, holding a spoon and a can, and looking directly into the camera. A very different group photograph shows him with Meinhof and Doegen (plate 5).

The photograph shown in plate 5 is in the collection of the Deutsches Historisches Museum in Berlin. Although it is included in Wilhelm Doegen's papers at the museum, he did not take the photograph himself, nor did he publish it in his writings. It is likely that the photograph, among other group photographs in a similar style, was commissioned by the art historian Adolph Goldschmidt. In the image, Mohamed Nur is seated second from left, with Carl Meinhof in the middle and Wilhelm Doegen at far right. So far, I have not been able to identify any of the other men seated at the table; they are identified as "East Africans" in the description provided by the archive. The exceptional elegance of this composition is related to Goldschmidt's staging of several of the group photographs to mirror scenes in Dutch paintings of the seventeenth century.[18] Composed in this specific way, the quiet scene with men in suits seeks to represent the honor of knowledge production. The reality of the camp does not surface in the picture. The seating arrangement around the small table places Meinhof at the center of the photograph, giving him just enough space to create a sense of distinction and allowing only his (white) hands to rest on the table. The composition produces an almost tangible aura around the central figure—Meinhof the scholar—and thus seems to present Doegen and the internees as his disciples.

Reassembling archival objects related to Mohamed Nur's presence exposes the efficacy of the operations of coloniality in epistemological as well as in artistic practice; it utilized the presence of POWs as exotic bodies and as embodiments of the knowledge desired by the German linguists.[19] Nur's archival shadow is filtered through a variety of practices and aims, which constitute the *Verwertungsmaschine* of the colonial society in Germany in the prewar and war years. For this purpose, I translate *Verwertungsmaschine* as "apparatus of utilization," a term that captures the objectified role assigned to the black body which, as Stoecker (2013, 73) writes, was seen as fundamental even for African language assistants at the institutes where African languages were taught in Germany. Visualization—but also a wider sense of auscultation, a listening to languages through the body of an informant who was prompted to speak into a recording device—created a plethora of contradictory documents.[20] Motivated by both scientific and artistic interests, linguists, phonologists, and others—including directors of *Völkerschauen* and cabarets, artists, and photographers—produced visual and sound documents for the *Verwertungsmaschine*, which then were also used as propaganda against the colonial soldiers of the Triple Entente.

The *Völkerschauen* as Camps and POW Camps as *Völkerschauen*

The term *Völkerschau* describes a kind of show that mostly took place at zoos or fairs, in which performers defined as "foreign" or "other" were presented on stage together with props, costumes, and animals, for a paying audience. *Völkerschauen* responded to and shaped European fantasies of "the exotic" and "the savage" by presenting non-European people to a larger German public. I do not use the often-employed term *human zoo* to describe these shows because its implication of captive performers does not accurately reflect the situation of ensemble members who may have joined such companies by choice, even if under duress (as in Nur's case) and even if the circumstances under which they would be performing in Germany may not have been made clear up front.

Völkerschauen in Europe shaped metropolitan audiences' practices of looking, popularizing notions of racial difference and practices of objectifying non-European people, who were exhibited in line with various, always racializing, depictions of "exotic peoples" and their imagined lifestyles. Performers danced, played music, mimed fights, and sometimes even reenacted scenes from colonial wars, which were choreographed to be presented at such shows (Pfäffinger 2015, 11). Performers were photographed and drawn by artists and examined and measured by anthropologists, who also made plaster or wax casts of performers' heads and faces (Zedelmaier 2007, 192). Life casts that were produced with ensemble members of *Völkerschauen* were shown at Castan's Panopticum in Berlin, and in so-called war exhibitions in Vienna in 1915 and 1916. These shows, together with the parasitic practices of knowledge production they facilitated, may sound bizarre now, yet they were a mass phenomenon in European cities in the late nineteenth and early twentieth centuries. Between the 1870s and the 1930s, around four hundred such *Völkerschauen* toured in Germany, attracting millions of visitors (Zedelmaier 2007, 186). The exhibition called *Nubier* (Nubians), which ran at the Berlin Zoo in 1878, attracted around sixty thousand visitors every Sunday. Carl Hagenbeck's *Ceylon Show* was seen by one million visitors during its two-and-a-half-month run in Paris in 1886. A total of 103 people from German colonies who found themselves exposed at the *Gewerbeschau* (trade show) in Berlin Treptow were stared at by two million visitors within seven months (Pfäffinger 2015, 326; Zedelmaier 2007, 199). This means it is at least statistically possible that every citizen of Berlin, which had fewer than two million inhabitants at the time, had seen the show in Treptow.

Hilke Thode-Aurora (1989, 31) writes that Somali shows were particularly popular in Germany (see plate 8). The Hamburg Zoo's owner, Carl Hagenbeck, who was also a preeminent director of such shows in Germany, advertised visits to the so-called Somali Village he presented as an opportunity to see Africa without traveling there.

> Um Afrika zu sehen macht man keine lange Reise, sondern geht zu den 100 Somalis im Somalidorf.[21]

> To see Africa one needn't make a long trip, just go to see the 100 Somalis in the Somali Village.

Several of these Somali Dörfer (Somali Villages) that toured Germany comprised more than one hundred performers. The continuation and success of these Somali shows over a period of thirty years appears to be the result of a collaboration between Carl Hagenbeck and Somali businessman Hirsi Egeh Gorseh, who, like Mohamed Nur, came from Berbera. Thode-Aurora (1989, 37) writes that Hirsi Egeh Gorseh recruited performers for the shows and exported German sewing machines to Somalia.

While performers signed contracts to perform in Hagenbeck's shows, power relationships were clearly asymmetrical, and the ensembles lived under the tight control of the shows' directors. Performers were often confined to the fairgrounds, and, apart from the general violence of voyeurism and objectification that was intrinsic to the shows, physical abuse also occurred. In some cases, the private lives—and deaths—of the performers were shamelessly exploited to advertise the shows. This was the case with the funeral of a woman who performed in an ensemble called Dahomey-Amazonen (Amazons of Dahomey). After she died of pneumonia in Munich in 1892, her funeral was staged as a public event (Dreesbach 2005, 120). While performers may not have accepted without resistance the regime they were subjected to in the shows, historical documentation of mutiny is scarce. If it surfaces at all, it mostly filters through in the comments of European observers. An example of such contemporary commentary is the irritated remark of the German anthropologist Felix von Luschan that the group of Ovaherero, who were part of an ensemble in Treptow in 1896, insisted on wearing suits instead of the traditional costumes he had imagined them wearing.[22] In 1910 the Berlin *Tägliche Rundschau* reported a rough encounter between performers and the Luna Park showground's guards on August 27, when a group of young men wanted to leave the Somali Dorf in Berlin Halensee (see plate 7) but were prevented from doing so by the guards (Mergenthaler 2007, 52). This incident speaks of constraints on freedom of movement of the performers who may have found themselves in a camp situation on the fairgrounds.

Historical records of the viewpoints of performers are almost nonexistent, writes Thode-Aurora (1989). Apart from a translation of the diary of a performer from Labrador, Abraham Ulrikab (or Ulrikeb), and the letters of Friedrich Maharero from Namibia, no written documentation seems to have surfaced that conveys the views of the performers on their experiences in such shows (Thode-Aurora 1989, 36; Ulrikab et al. 2005).[23] Here, too, the colonial archive directs (later) research and barely seems to allow the appearance of the perspective of performers. The shows' focus on the visual has created a particular archival trail that is fixated on seeing others in predefined ways.

Mostly this means that what appears in photographs, postcards, and in film is the staging of those deemed exotic.[24] Among the postcards, which mostly repeat the motifs of the shows, only a few photographs depict the visitors (see plate 7).

As far as I know, no photographs taken by the performers themselves are known to exist. The available archival collections with documentation on *Völkerschauen* in Germany speak of practices of visualization, of staging, of display, and of men and women being in Germany to be looked at. A plethora of photographs, posters, and postcards, but also silent movies, which show staged fights or men and women with camels and props, perpetuate this gesture of exhibition (see plate 8).

None of the studies on the histories of the *Völkerschauen* I have seen have used acoustic traces or oral archives. Yet these do exist. After the opening of my exhibition *Der Krieg und die Grammatik: Ton- und Bildspuren aus dem Kolonialarchiv* (War and grammar: Audiovisual traces from the colonial archive) at the MARKK, Hamburg (October 22, 2019–February 23, 2020), I received an email from Samatar Hirsi Egeh, the great-great-grandson of Hirsi Egeh Gorseh from Berbera, who had organized the *Völkerschauen* with Hagenbeck. Samatar Hirsi Egeh had been living in Germany for ten years. The story of his great-great-grandfather's journey and work in Germany still exists in the memory of his family. This includes remembering Mohamed Nur as the teacher of Hirsi Egeh Gorseh's children in Germany.

Acoustic echoes of *Völkerschauen* are also available in German sound archives. As mentioned in chapter 1, the musicologists at the Berlin Phonogramm-Archiv took the presence of such shows at the city fairgrounds as an opportunity to make recordings. As part of the documentation of a recording with Somali performers in Halensee, in Berlin, the archive includes a letter from the organizer of the show, a Mr. Holz, responding to a request from the Phonogramm-Archiv for permission to record the performers:

HALENSEE 8.6.1

LUNA PARK, SOMALI DORF

An das
Phonogramm-Archiv des
Psycholog. Instituts der Univ. z. Berlin
Berlin NW
Dorotheenstr. 95/96

Unter höfl. Beantwortung lhres verehrten Briefes vom 6. d. Mts. teilen wir lhnen hierdurch mit, daß es uns sehr angenehm ist, Sie zwecks

Aufnahme von Gesängen etc. hier im Somali Dorf begrüßen zu können.

Jedoch möchten wir Sie bitten, uns vorher Zeit und Stunde anzugeben, wann Sie hier eintreffen wollen, wenn möglich zwischen 11–12 Uhr vormittags.

Hochachtungsvoll

Continental Enterprises Ltr.
E. Holz[25]

In the letter above, the director of the so-called Somali Dorf in Halensee confirms that they agree to a visit from the researchers with the purpose of recording songs and ask that an appointment be made in advance, preferably for a recording session in the morning. The result of that recording session is a collection of twenty-four now-digitized wax cylinder recordings at the Berlin Phonogramm-Archiv, produced at Luna Park in 1910.[26] In 2018, Bodhari Warsame translated one of these recordings, in which a man called Shire Rooble sings of his reasons for coming to Berlin:

Barliin waxaan u imid
Baahi lacageede,
Qawi lagu ma helo quud
Hadduu Eebbe kuu qorine,
I gargaara qayrkayow
Mar bay ni qabsan doontaaye.

Berlin I came because of
The need for money!
Your own efforts cannot always secure a fortune!
If God does not permit it to you.
Oh my peers help me
For you too might one day be in the same situation![27]

Taken on their own, or together with Mohamed Nur's comment on the situation of travelers (to which I will return later in this chapter), these recordings shift the overall theme of the archival trail: the performers do not speak of being stared at, nor do they address the voyeurism of the audiences. Instead, they focus on themselves and express their own reasons for traveling to Germany in the first decades of the twentieth century. Warsame's translation bursts the bubble of exotic otherness: the performers' reasons for embarking on the journey across the sea are no different from those that many migrants in the past and in the present give: to make a living. Even as small fragments, these recordings are able to shift the perspective on the shows: the performers are heard as speakers; Rooble represents them as migrant laborers instead of as objects or victims of European voyeurism. While even a critical review of the images of the *Völkerschauen* cannot go beyond the visual repertoire of the shows, recordings convey the positions of their speakers in their own voices. A systematic translation and study of all recordings could provide archival documents that would allow researchers to break the durable spell of the spectacle.

Völkerschauen were popular in Germany until well into the 1930s. The fact that non-European soldiers in the Halbmondlager in Wünsdorf were exposed to the voyeurism of visitors and the attentions of anthropologists is arguably one of the effects of the massive success of these shows. As described in chapter 1, linguists and anthropologists were alerted to the presence of the non-European soldiers in the internment camps who subsequently became the objects of their research.[28] Margit Berner (2010, 235) and Britta Lange (2010, 313) have described the methods used to take measurements of prisoners as well as the application and development of visual methodologies, together with the impact of nationalism that also led to the imagination and documentation of an alleged variety of racial differences among Europeans. Andrew Evans (2010a, 201; 2010b) writes of these practices as resulting from a specific combination of intensified nationalism during World War I and the circumstances of research in the POW camps, which were not too far removed from the research methods practiced under colonial circumstances. In fact, several of the researchers involved with the KPPK had previously conducted anthropometric examinations of people in colonized areas or on colonial expeditions (for instance, to the South Pacific) and both von Luschan and Pöch had produced a combination of measurements, life casts, photographs, and sound recordings in southern Africa (Hoffmann 2020a). The research activities that took place in the camps, and the visits of photographers and artists to the camps, led to the accumulation of a vast quantity of written and visual documents, both published and unpublished.[29] The immense scale of data production is exemplified in the anthropometric measurements of 1,784 prisoners of war between 1916 and 1917 in Germany, which Egon Eickstedt had amassed under von Luschan's supervision (Lange 2010, 318). The lists of measurements, the photographs, life casts, and also the recordings form a material trace of the operation of anthropologists, linguists, and musicologists in the POW camps of Germany and Austria. These visual, written and sonic materials also became the means of production for the careers of several of the academics involved (Schasiepen 2021; Eickstedt 1921; Weninger 1927). Unsurprisingly, nationalist sentiments that informed the research of the KPPK bled into propagandistic publications (Koller 2011a, 2011b; Evans 2003; Stiehl 1916; Doegen 1925, 1941). Although there were various agendas at play—for instance, in the studies of the anthropologists Felix von Luschan (1922), Egon Eickstedt (1921), and Rudolf Pöch (Weninger 1927); the artistic-anthropological portraits of Hermann Struck (von Luschan and Struck 1917); the photographs of Otto Stiehl (1916); the publications of Otto Reche (Evans 2003); and Leo Frobenius's (1916) strident description of the British army as "a tamer in a circus of peoples" ("Dompteur im Völkerzirkus")—the production of all these texts, photographs, and portraits was triggered by the availability of a racialized Other in the camps, albeit to different extents and ends (Frobenius 1916; Evans 2003; von Luschan and Struck 1917). Käthe Kollwitz, a renowned German artist, wrote in her diary:

> Viel kleine elende Menschen. Hin und wieder auch schön und groß gewachsene. Im ganzen und großen wirkt diese Anhäufung gefangener Feinde deprimierend. Es erinnert etwas an Hagenbeck (über Wünsdorf, 29. September 1914).[30]
>
> Many small, wretched people. Sometimes also beautiful and tall people. Altogether this accumulation of captured enemies is depressing. It reminds me of Hagenbeck (about Wünsdorf, September 29, 1914).

While Kollwitz did not refer to the camps as an unprecedented *Völkerschau* and an opportunity, as the Austrian anthropologist Rudolf Pöch did, she too compared the camp to a *Völkerschau* by referring to Carl Hagenbeck. The iterated reference to some camps as *Völkerschauen*, the practice of examining performers and prisoners alike, the circulation of postcards with images of non-European prisoners—much in the same manner that postcards of *Völkerschauen* had been distributed before the war—all suggest that the practice of showing "exotic" people in confined spaces may have directly impacted the way the imprisoned soldiers were seen in Germany (Lange 2010). By the time the POWs were interned in German camps, *Völkerschauen* had been around for at least eighty years.[31]

From the Show to the Camp

From the texts in von Tiling's (1925) publication, we know that Mohamed Nur traveled to Germany with a *Völkerschau* company in 1910 or 1911. He had not refused the embrace of the *Verwertungsmaschine*, which seized him before he even reached the *Kaiserreich*, and which he, in turn, had used to leave what I read as an untenable situation in Aden in 1910. What I imagine as the "apparatus of utilization" was, at least in this case, a kraken with no single mind, no exact direction, yet operating according to the logic of coloniality that saw people of color as objects of and for knowledge production, for the creation of exotic images, and as objects for the curiosity of public audiences, scholars, and artists. Between 1911 and 1922, Nur's presence in Germany was exploited in several ways, some of which were useful to Nur himself, others of which were not.

Some of the visual traces resulting from Nur's involvement with the *Völkerschau* illustrate the travelogue he dictated to von Tiling. The postcard showing the school at the Somali Village in Luna Park (plate 9) does not name Nur (standing at right), but it does connect to his account in *Somali Texte*. Here, a grammar of Somali contains snippets of a narrated journey, in which Nur explains how he came to Germany in 1911:

> After staying in Aden for 2 months already, a friend said to me: "Why don't you go with the traveling company (of performers) as a teacher?" I replied: "I do not want to go with them!" He said then, "Just go with them, there is no shame in this—and they only stay for a year," so I said, "All right, fine" and went with them and taught the children. (von Tiling 1925, 52)

A quarrel with the director of the company left Nur stranded in Germany. Interestingly, Nur did not articulate the cause of the quarrel himself; instead, von Tiling (1925, 52) added to his narrative, in parentheses: "Explanation: the director of the company had assured M. Nur that he would never have to participate in the dances and fighting performances. When he did request that he participate in the performances, the conflict came about." What else Nur told von Tiling about his engagement at the *Völkerschau*, apart from the narratives she wrote down, which then became the basis of the explanation she offered, remains unclear. This intervention in parentheses, the superimposition of von Tiling's explanation onto Nur's text, demonstrates Nur and von Tiling's complicated coauthorship of the publication. A pamphlet for a Somali Village shown in Vienna, which displays the photograph shown in plate 9, suggests the source of conflict:

PLATE 1 Stereo photograph of Abdoulaye Niang, undated. Department of Evolutionary Anthropology, University of Vienna.

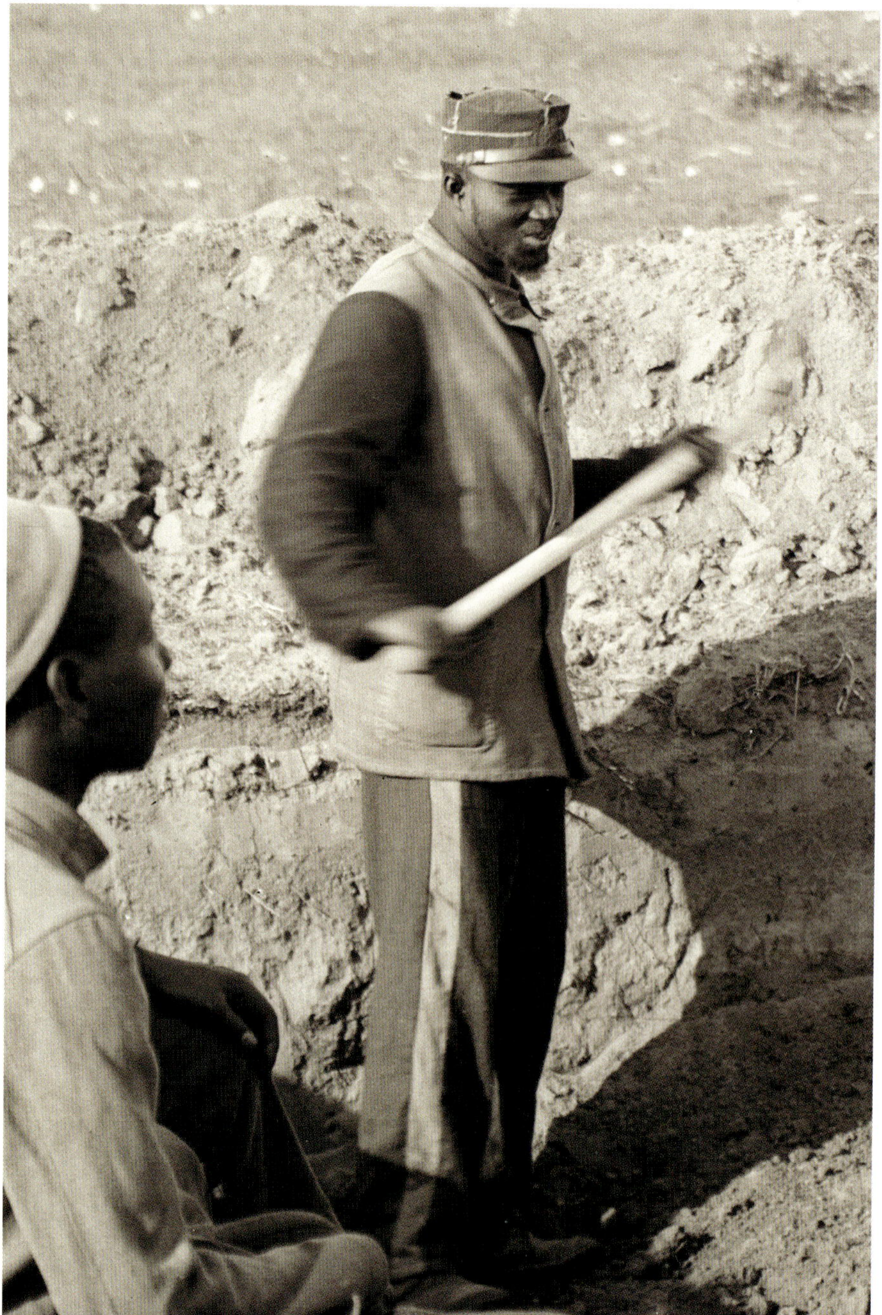

PLATE 2 (*opposite*) Abdoulaye Niang in the POW camp at Turnu Măgurele, Romania, undated. Detail of a group photograph, probably taken by Leo Frobenius. Frobenius Institut Frankfurt.

PLATE 3 (*above*) POWs from Madagascar and Comoros in the camp at Wünsdorf, Germany. Asmani ben Ahmad is second from left, the linguist Paul Hambruch third from right. Photographer unknown, undated. From Doegen 1941, 64.

PLATE 4 Inhabitants of barrack 13 at Ruhleben camp, which was designated for black internees. Mohamed Nur is third from right, front row, crouching. Photograph taken by Albert Grohs, undated. Maurice L. Ettinghausen Collection of Ruhleben Civilian Internment Camp Papers, Harvard Law School Library, Historical and Special Collections.

PLATE 5 Staged meeting (?) at Ruhleben camp, with Mohamed Nur second from left, linguist Carl Meinhof in the middle, and philologist Wilhelm Doegen at far right. Photograph by Adolph Goldschmidt, undated. Deutsches Historisches Museum, Berlin, Nachlass Wilhelm Doegen.

Rudolf Zeller pinxit A. Claussen phot.

Muhammed Nur

PLATE 6 (*opposite*) Rudolph Jacob Zeller, *Mohamed Nur*, 1921. From von Tiling 1925, 1.

PLATE 7 (*above*) Entrance of "Somali Village," Luna Park, Halensee, Berlin. Postcard, 1910/11. Private collection of Clemens Radauer.

PLATE 8 Somali Show with visitors, 1927.
Photograph no. 14/22. Private collection of
Clemens Radauer.

PLATE 9 School in "Somali Village," Luna Park, Halensee, Berlin. Mohamed Nur stands at right. Postcard, probably 1912. Private collection of Bodhari Warsame.

PLATE 10 (*below*) Top left: Max Slevogt, *Der Sieger* (The victor) (detail), 1912, Kunstpalast, Düsseldorf. Top right: Max Slevogt, *Brustbildnis des Somali Hassanó* (Bust of the Somali Hassanó) (detail), 1912, Kunstpalast, Düsseldorf. Bottom: Max Slevogt, *Hockender Afrikaner* (Crouching African) (detail), undated, private collection.

PLATE 11 (*opposite*) Albert Kudjabo in the POW camp at Soltau, near Münster, Germany, with a Melanesian "speaking drum." *Berliner Illustrierte Wochenschau* 9 (33), 1924.

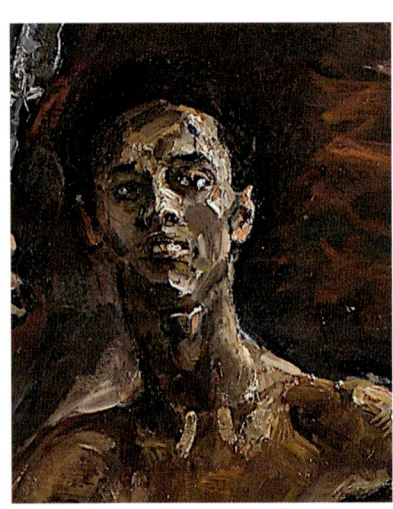

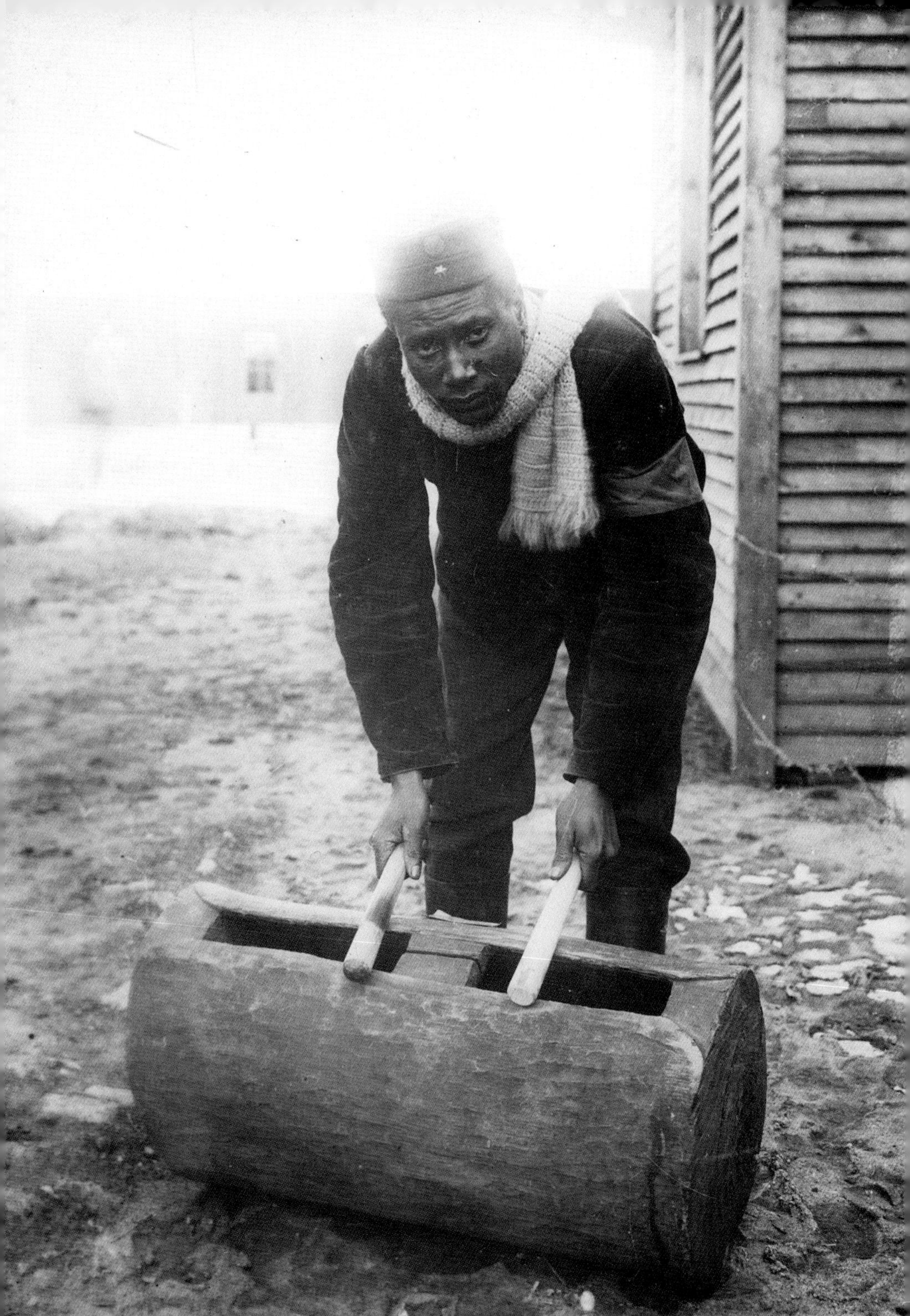

PLATE 12 "Gold Mine at Kilo-Moto" (Democratic Republic of Congo). Postcard, 1926. Courtesy of the author.

PLATE 13 Albert Kudjabo, 1916.
Courtesy of Odette Kudjabo.

27.2:2 b

PLATE 14 (*opposite*) Family photograph with Albert Kudjabo. Courtesy of Odette Kudjabo.

PLATE 15 (*below*) "Die Fetischhöhle [the Fetish Cave] in Togo-Kratschi" (Ghana). Photograph by Alexander von Hirschfeld, ca. 1927. MARKK, Hamburg.

PLATE 16 Sound/film installation at the exhibition *Der Krieg und die Grammatik: Ton- und Bildspuren aus dem Kolonialarchiv* (War and grammar: Audiovisual traces from the colonial archive), at the MARKK, Hamburg, 2019. Courtesy of the author.

Fortwährend von dichten Scharen umlagert ist auch die Schule wirklich ein malerisches Bild der lebendigen, frischen, afrikanischen Jugend . . . die Kinder genießen elementaren Unterricht beim Mullah, werden von ihm in die Geheimnisse des Korans eingeweiht.[32]

Constantly surrounded by a dense crowd of visitors, the school is a striking picture of the lively, vital African youth . . . the children enjoy elementary instruction by the mullah and are initiated into the mysteries of the Koran.

The mullah, or religious teacher, at least in the photograph and in the oral family archive of Samatar Hirsi Egeh, great-great-grandson of Hirsi Egeh Gorseh, was Mohamed Nur. Nur probably experienced the showcasing of his work and the education of the children in the Somali Village as part of a spectacle. Boundaries between performance and the private life of performers were clearly not respected.

Nur's Archival Echo: Fragments of Somali Poetry, the "Mad Mullah," and Some Remarks on Migration

Perhaps the most opaque but also the most intriguing aspect of Nur's archival trace in Germany is the echo his voice has left in the records of the KPPK. The recordings produced with Nur, held at the Berlin Lautarchiv, were recorded in Ruhleben by Carl Meinhof and Wilhelm Doegen on April 27, 1917.[33] Nur's recordings, like many other historical recordings that contain spoken and sung texts, include fragments of particular oral archives and discursive formations, in this case, of Somali oral poetry. The recordings are double-storied, doubly inscribed archival objects: they are part of an archive of colonial knowledge production and thus speak of their making as much as of the interests of the researchers. Yet they also communicate aspects of other practices of knowledge production, forms of transmission and storage from a different archive and intellectual tradition. The German linguists did not identify these aspects in 1917, nor did they do so in the one-hundred-year archival life of the recordings since then. One of the reasons for this neglect of genres and content is the fact that the archival organization of the Lautarchiv systematically obscures the double inscription of these acoustic documents in its registers. The suspension of communication that took place at the moment of knowledge production extends into the archive's register, which indexes only languages and assumed ethnic groups, not spoken or sung textual content. By means of recording and archiving, snippets of discourse became complete specimens that stood for entire languages. In the written documentation, the semantic meaning and provenance of these textual fragments were left out; yet the recordings have acoustically conserved some of the content that was once deemed irrelevant. In one of the recording sessions in Ruhleben, Mohamed Nur tapped into a vast archive of Somali oral poetry. This archive of orally recited and transmitted poetry has played a major role in political life in Somalia and has long been treated as a documentation of Somali history (Samatar 1982; Andrzejewski 2011, 6). Engaging with these recordings, we thus enter the repository of knowledge of Somali orature, which I can approach only in dialogue with Bodhari Warsame, based on his translations and together with analyses of its forms, genres, and political impact that have been made accessible to Western readers in the body of scholarly work on Somali poetry.

In his monograph on oral poetry and Somali nationalism, Said Sheikh Samatar describes the function and social status of Somali oral poetry as follows:

> Oral verse indeed seems to embrace a wide range of cultural and material activities in pastoral life. To begin with, poetry is not the craft of an esoteric group of beauty-minded men whose role in society is at best marginal. On the contrary, his craft places the Somali poet in the mainstream of society and his energies and imagination are constantly drawn upon for social purposes: his kinsmen expect the Somali poet to defend their rights in clan disputes, to defend their honor and prestige against the attacks of rival poets, to immortalize their fame and to act on the whole as a spokesman for them. . . .
>
> While the cultivation of beauty and high thought, the perfection of language and imagination and other virtues associated with art are the tasks of a Somali poet, his paramount responsibility is utilitarian: to inform, persuade, or convince a body of kinsmen of the merits of whatever task he seeks to undertake. (1982, 56)

Highly formalized poetry was thus the main mode of negotiation between clans, which were not ruled by higher authorities. Verses were memorized to communicate messages transmitted over vast distances. Poetry was the sophisticated medium of negotiation for contracts, marriages, and debates about resources. Genres of poetic speech were used to wage wars and to terminate them; poetry was "the principal vehicle of political power" (Samatar 1982, 56). Poets acted as spokespersons, and clan leadership was often achieved via poetic skill (Andrzejewski 2011, 6).

Somali poetry entails distinct genres and subgenres. The genre name *maanso* describes several subgenres of poetry that articulate social and political matters (Orwin 2005, 279; Samatar 1982, 74). The verses of this genre are (and were) composed prior to performance and must be presented verbatim in every performance by their composer or by a performer who has memorized the verses (Samatar 1982, 57). This differs from, for instance, oral poetry in Otjiherero (in Namibia and Botswana) or in isiXhosa (in South Africa), which occupied, and in some aspects continues to occupy, a similar position in the political life of the societies of the speakers of Otjiherero or isiXhosa. Unlike performances in Otjiherero, for instance, *maanso* poetry does not encourage or allow individual versioning of the verses. Instead, it values and recognizes authorship as ownership of verses and protects the author's rights by denouncing any kind of borrowing or plagiarism. The strict formalization, the rules of alliteration and meter, together with a concept of ownership led to very stable memorization, or what Martin Orwin (2005) has called "definitive texts." These texts respond to other verses and poets but do not bleed into each other in cut-and-paste practices (as in Otjiherero poetry; see Hoffmann 2015). Long debates or disputes fought out in verse are called "chains of verses" and include multiple poets' verses, sometimes continuing over decades and being named and recalled as specific intellectual discourses and engagements (Barnes 2006).

Nur's performances for the recording session presented fragments of such resonating, corresponding swarms of voiced texts. As fragmented archival objects, the recordings have lost much of their resonance. What the Lautarchiv has kept may have been perfect specimens of language for the linguists, but the recordings are also ossified acoustic fragments that may or may not resonate with an oral archive in the present. The fact that African genres of record keeping were (and are) rarely identified in the registers of acoustic collections of the Lautarchiv or in other sound archives, together with the fact that most of the content and genres of the speech and songs recorded were not specified, gravely hinders the identification of specific texts or content as historical sources when searching archival registers, for instance. As a result, the translation of "example sentences" or "traditional songs" often allows for the resurfacing of unexpected texts and content. In the case of Mohamed Nur's recordings, some aspects speak of an episode in colonial politics that found its way into Somali historiology and historiography and that corresponds to Nur's narratives that appear in von Tiling's text.

File number PK 860/1 registers the first of five recordings produced in Ruhleben with Mohamed Nur. The title Wilhelm Doegen noted for the song

Nur sang into the funnel of the recording device was "Spottlied auf den tollen Mollah" ("Mocking song about the mad Mollah"). As with most of the recordings, the song was filed, archived, and probably never listened to again for the past one hundred years.[34] Bodhari Warsame's first translation of the recording in 2013 revived fragments of a poetic battle that took place in Somalia (which was then British Somaliland) in the years before Nur left the country. Beginning with a short chant without words that sets the tone and the rhythm, Nur sings:

Hoyaalayay, hooyalay, hooyaalayay-hooye
Hoyaalayay, hooyalay, hooyaalayay-hooye
Hoyaalayay, hooyalay, hooyaalayay-hooye

Toolmoonihii Maxammad baw tegay Ilaaheenee
Tii uu na faray looxayaday noogu taal waliye
Safar waa la taakulin jiree la ma tayiisayne
Tii uu na faray looxayaday noogu taal waliye
Ma Talyaanigaasaa Mahdiya tanuna waa yaabe!
War maxaad ku gabyaysaa? War gabaygaasi waa gabaygii
Ina Cabdille Xasan weeye.
Isagu daallinkii carruurtii gawrgawracay ee wixii Ilaahay yiri isagu diiday.
War kaa waxaynnu doonaynnaa in aynnu soo qabanno. Inaynnu halkaa ka saarno, haddii Ilaahay . . . Ilaahay . . . Waa in la soo qabtaa.
Waxaynu doonaynnaa in aynnu qabanno oo soo xirno.
War isagu kolkii hore Muslim buu ahaa. War kolkii hore Muslim buu ahaa, imminka se waa riddoobay.
Daallin buu noqday, carruurtii buu gawracay, maqashii buu ka qaaday, geelii buu qaaday, arigii buu qaaday, xoolihiiyoo dhan buu dadkii ka nahaabay. Isagu nin nahaaboo kibir weeye.
Inaynnu ninkaa la dagaallannaa baynnu doonaynnaa. War . . . isagu waa tahay.
Berri waa inoo col. Berri waa inaynnu ku guurro.
Hadalku sidaa weeye.
Guddoon!

The great messenger has met Allah
His instructions are still written on our religious tablets [*looxayaday*]
Traditionally, travelers used to be assisted, not troubled

[next line incomprehensible]
the instructions of the great messenger are indeed still written on our tablets
And now, by all wonders
Does that Italian heathen now claim to be a savior [*Mahdiya*]?

What are you talking about?[35]
This is indeed a poem mocking 'Abdille Hassan
The violent transgressor who murdered women and children at will, and rejected the word of Allah.
We want to catch him and chase him away from the land.
He was once a true Muslim but has now proved to be an apostate.
He became a transgressor and murderer of children, stealing animals, looting camels, looting goats.
He is a swindler and is arrogant.
We want to fight this man, that is our decision!
Tomorrow is the battle, tomorrow we attack!
That is agreed![36]

Mad Mullah (or Wadaad Waal) may not have been a name Wilhelm Doegen or Carl Meinhof recognized when they recorded these songs with Nur in Ruhleben. Perhaps because of its spelling (Mollah instead of Mullah), the title of Nur's recorded performance lost all connection to historical characters and events in the first decade of the twentieth century in Somalia. The derogatory name Mad Mullah was a characterization given to Mohamed 'Abdille Hassan (1864–1920), first in the region of Berbera, and then subsequently popularized by his main antagonist, the poet Ali Jama Habil, in a long political dispute fought out in orature. The British colonial power picked up this name (Mad Mullah) quickly and eagerly as an early instance of the articulation of (Muslim) religious fervor, nationalism, and resistance against colonialism that was to signify, or indeed construct, a concept of unique madness connected to political Islam. Samatar (1982, 184) suggests that the adjective *waalan*, which is mostly translated as "mad," did not necessarily designate insanity in Somali but could instead have pointed to something extraordinary, perhaps excessive, as in "madly brave." In the 1970s, writes Samatar, the rise of nationalistic historiography in Somalia discursively transformed Mohamed 'Abdille Hassan from (Mad) Mullah into honored Sayyid, a true descendant of the prophet (183).

Mohamed 'Abdille Hassan (the name is also written as Mohammed Abdullah Hassan or Maxamed Cabdulle Xasan) was an Islamic scholar, a famous poet, and a member of the Salihiyya Sufi order, which originated in Sudan. Around 1895 he settled briefly in the port city of Berbera, which at that time was also the administrative center of British Somaliland (Hoehne 2016, 1). From there he agitated—also in verse—against cooperation with the British colonizers. With up to five thousand followers, Mohamed 'Abdille Hassan sought to create an autonomous space and an alternative religious order (Abbink 2003, 341). From 1899 onward, he led a militant insurgency against Christian invaders; it became known as the Dervish movement. His raids and aggressive military campaigns were also directed against Somalis who did not fully support the movement or follow its strict rules. This led to widespread communal repudiation of his movement, for instance, in the area of Berbera, where Mohamed Nur came from. In the texts published by von Tiling (1925), Mohamed Nur writes:

> When the priest started the rebellion, he wanted all Muslims to follow him in this holy war and chase the British out of the country.
>
> Initially, people followed him, but then they [his followers] did bad things. The people looted and robbed [people] and said: those who do not pray will be punished! Then people who had not prayed in their lifetime started to pray and turned their faces to the sky.
>
> Later, his followers left him. Then he did not distinguish between the British and the other people. . . . and some people who had never prayed in their lives and who did not know god's commandments, wrapped their heads with turban and said: we are dervishes. . . .
>
> Then the priest/teacher moved to Nugal and every year their troops went to war; they even killed scholars. And the priest and the poets sang against each other and abused each other terribly. (45–47)

Nur's second recording with the KPPK (PK 860/2) connects to his account of the Dervishes' looting and robbery of people in the area of Berbera:

Geerarrkaygu Xariiriyo
Waa xaqwaysanahaye
Xaakinkii na abuurow
Xaakinkii na abuurow
Xaqeennii luminaynee
Xigaalow nala kacooy!

Waxaan gaajo la xoodmee
Habeen noolba Xanjaadkiyo
Xundhurtay isku hayaaba
Hadduu Geela Xayaable xiksimaysanayaaye
Waa bishay manjaxaabo
Xagga noogu lagdeene
Xigaaloow nala kacooy!
Qabiilkaan xil dhibaynee
Xoolihiisa tabaynnin
Waa xayska Maqleede
Xigaaloow nala kacooy!
Haddaydaan hadalkaygiyo
Himmaddayda maqlaynin
Waa habeenkii madoowe
Waa halkaa iska daayoo!

My *geeraar*[37] is precious as silk!
I am violated!
Oh the All-just, our Creator will not forsake our property and pleas for help!
So kinsmen, join us!
Starvation bends my body
each and every night my backbone and stomach meet in painful hunger
Now Hayaable is milking my camels
It is the anniversary of the month in which they defeated us
Oh kinsmen join us!
A careless clan that does not strive to recover its property is like the brief rain at Maqleed
Oh kinsmen, come and fight with us!
If you don't listen to me, let it be like this![38]

The verses communicate the grievances of a man who has lost his herd, yet who also seeks to convince clansmen to fight against the raiders. While the recorded verses come without explanation, they correspond with the biographical notes that Nur narrated for von Tiling. Read together, the recordings position Nur in the middle of a political crisis in Somalia and contribute to our understanding Nur's position vis-à-vis the Dervish movement and his engagement in political intrigue. In *Somali Texte* (von Tiling, 1925, 62–92), Nur presents in detail an account of traveling to Mecca with the delegation under 'Abdallah Shihiri, a former British naval interpreter and former close ally of Mohamed 'Abdille Hassan. Nur mentions that the journey to Mecca with Shihiri's delegation was supported by British and Italian colonial powers.

In Mecca the delegation met the founder and head of the Salihiyya brotherhood, Sheik Mohamed Salih Rachid, to denounce the religious leadership of Mohamed 'Abdille Hassan. One of the scholars who traveled with the delegation, Ismaa'iil Isahaaq from Berbera, may have been young Mohamed Nur's teacher.[39] A year later, the delegation returned with a letter from Sheik Mohamed Salih Rachid, which "virtually excommunicated the Sayyid [Mohamed 'Abdille Hassan]" and questioned his religious integrity (Samatar 1982, 127). This condemnation echoes in Nur's recitation:

> He was once a true Muslim but has now proved to be an apostate.
> He became a transgressor and murderer of children, stealing animals, looting camels, looting goats.

The outcome of the campaign was Mohamed 'Abdille Hassan's exclusion from the brotherhood.[40] The upheaval that followed is characterized by the designation of the period of 1911–12 in Somali oral records as "the year of eating filth" (Abbink 2003, 344). By then, Nur had already arrived in Germany. After traveling with the delegation to Mecca, Nur, according to his biographical account, first returned to Berbera, the region he called home; and from there he went to Aden. In Aden he joined the company with which he traveled to Germany. Five years later, Nur found himself detained in an internment camp near Berlin. Much of what he spoke into the gramophone funnel, and which then, by the magic of disciplinary protocol and the power of linguistics as a field, became language examples, was related to this intrigue.

How Nur came to speak of this crisis in the moment of linguistic recording remains unclear. If there was a request that led to his recital of poetry, it does not appear in the archive. Was Nur, like Josef Ntwanumbi, who was also recorded in Ruhleben, perhaps prompted to present "traditional" texts, and therefore chose to perform fragments of Somali poetry?

Another possibility may be that Nur was responding to the spoken text of Ali bin Bedja, who was recorded minutes before Nur, and who spoke about being interned in Ruhleben. Could bin Bedja's recording have triggered Nur's reminiscence of how he got to Ruhleben in the first place? His anger against Mohamed 'Abdille Hassan is audible. The crisis in Berbera may have pushed Nur to join the campaign against the Sayyid, which played into the hands of the

British colonial powers. It is likely that his involvement in the campaign led to Nur's decision to join the *Völkerschau* ensemble in Aden and travel to Germany.

In Nur's recording, echoes of a chain of poetic debates surface that speak of the upheaval the Dervish movement had brought to people in Somalia, for instance, with the abandonment of rules of hospitality. Yet with the recital of these lines, Nur may also have expressed the hardship he experienced in Germany when he was stranded without a passport, or money, or the ability to speak the language after his falling-out with the director of the show. Here, Nur seems to recite aspects of Ali Jama Habil's poem, which appears in Samatar (1982, 167) as follows:

> And caravans are given to the safety of Allah!
> But he wantonly cuts the tendons of weary travelers and gorges on
>   their dates![41]

Only fragments of Nur's recordings are congruent with poems I found elsewhere, but these respond to the debate between the Sayyid and his antagonist, Jama. The appearance of this connection may be related to the historical significance of that debate. Yet Nur's choice of material to recite also speaks of his preoccupation with the crisis in his country that led to his migration to Germany.

A.K.A. "Hassanó"

Nur's visual trace, found in texts on art history, in museums, and in auction houses under the pseudonym "Hassanó," could not be less related to his biography or his intellectual and political engagements in Germany and Somalia. Images painted and drawn in 1912 by the German artist Max Slevogt contrast sharply with the acoustic and written documents that Nur left in Hamburg and Berlin. According to the art historian Hans-Jürgen Imiela, Slevogt produced at least six images of Mohamed Nur in March 1912 (for three examples, see plate 10). However, I count nine representations of Nur, not all of which I can locate today (Imiela 1968, 137).[42] Nur's acting as an artist's model is related to the *Völkerschauen* in several ways. First, Nur's work as a performer must have led to his encounter with Slevogt, who regularly attended opera, theater, and dance performances in Berlin. Second, motifs from *Völkerschauen*, that is, from practices of staging and exoticizing people of color, are visible in the no less fantasizing and racializing artworks that

Slevogt created with Nur. Given that the paintings depict Nur as a naked dancer or clad in a loincloth, variously with spear, leopard skin, or feathers, it is not surprising that the Somali intellectual did not mention his work for Slevogt in the texts that appear in von Tiling's *Somali Texte*. Apart from a comment by a contemporary of Slevogt's, who describes Nur as a "dunkelhäutiger Herrenmensch,"[43] a short mention in Imiela's (1968, 137) biography of Slevogt, and a mention in Hugh Honour's (2012) contribution to *The Image of the Black in Western Art*, not much can be found on these images and their actual production. The absence of other representations, or rather, the missing connection to Nur's work as a teacher, religious instructor, and language assistant, facilitates Slevogt's fantasies that turned Mohamed Nur into an "Ashanti," a "Somali Warrior," and a "squatting African" and form an assembled image of a man who worked under the pseudonym of Hassanó.

How exactly Nur met Slevogt remains a point of speculation: according to Nur's narrative in von Tiling's text, he left the *Völkerschau* and found himself stranded in Germany without a passport or a work permit, or the ability to communicate in German. He fell on hard times when his money and watch were stolen by a man who had purported to help him. Nur visited the British and Turkish consulates in Hamburg and Berlin but could not obtain a passport and was therefore unable to work. He explains:

> Da wußte ich nicht, was ich tun sollte, den ich war ein Mann, der weder die Sprache kannte, noch etwas von den Leuten wußte! . . . Und die Polizei sagte zu mir: "wenn du keinen Ausweis hast, kannst du keine Arbeit bekommen." (von Tiling 1925, 53)

> Then I did not know what to do, because I was a man who neither spoke their language nor knew people there. And the police said to me: "if you do not have a passport, you cannot work."

While waiting for his passport to be sent to him by his brother in Berbera, Nur had limited choices to earn a living. He joined various ensembles, performing in shows, which he did not describe in his narrative. Traveling with one of the ensembles, he learned to play the cello. It must have been during this time that he also met the German artist Max Slevogt, who lived mostly in Berlin at the time.

Max Slevogt's artworks show a keen interest in performers and performances: he painted an opera singer and several dancers (Anna Pavlova was one of them), and he created illustrations and stage designs for theater and opera. If Slevogt met Nur at a show or performance, it would probably have been easier for Slevogt to work with him for several days (weeks?) than with a member of a *Völkerschau* ensemble, who would have been confined to the showgrounds. Yet Slevogt seems to project what he must have seen at Hagenbeck's shows onto the paintings he created with Nur: the white feather headdress that appears on Nur's head in *Der Tänzer* (The dancer) and in the drawing "Somali Warrior" is exactly the headdress one sees on postcards of *Isa-Krieger* (Isa warriors) from Hagenbeck's *Ethiopia Show* (1909), and Hagenbeck's *Galla Truppe* (1908). This feather headdress is not entirely fantastical. Bodhari Warsame states that single, white feathers were worn under certain conditions by men in Somalia. Wearing them signaled heroism or accomplishment in warfare. Clearly, this was not appropriate attire for a Muslim intellectual who had not fought in wars but had dedicated his life to scholarship. On the *Völkerschau* stage, however, the feather became an arbitrary, exchangeable symbol of exotic otherness. Slevogt's paintings transferred these fantastical styles of dress and representations with spears and leopard skins from the dubious sphere of the exotic spectacle directly into the heart of the art museum. As they became part of museum collections, these representations entered the visual canon of high culture, where until now they have remained largely unquestioned. At the time of my first inquiry at the Kunstpalast in Düsseldorf, where two of Slevogt's paintings reside, nothing but the pseudonym "Hassanó" was known about the model who posed for Slevogt. In 2010 the art historian Hugh Honour could find no racism in his review of Slevogt's paintings of Nur as a naked dancer with a feathered head, or a squatting "Ashanti." He writes:

> Instead of depicting him in the frozen postures of an academic life-class model, Slevogt concentrated on his movements and *recording his physique specifically as a Somali*—crouching, squatting, standing on one leg by a dead leopard, and dancing with a lance and a leopard skin—as for ethnographic illustration. In the subject picture, no less broadly handled, these objective records of visual impressions were transformed into an image of "savage" virile sexuality by the addition of three white women. . . . *racial conceptions barely perceptible* in the studies rise to the surface in the picture together with erotic fantasies. The studio model is returned to Africa, the theater of colonial warfare and the dark continent of European imagination. (Honour 2010, 243, emphasis added)

Being assigned a specifically "Somali physique" by the reviewer of the portraits in 2010 shows that Mohamed Nur's images have not escaped the essentializing spell of the *Völkerschau* even a century after Slevogt painted him. Interestingly, Honour's critique only begins with the appearance of three captive white women in what he sees as Slevogt's main composition: the larger painting called *Der Sieger* (The victor), in which Nur is depicted with a loincloth against the backdrop of a dark sky, fires, and wasteland, watching over the women. However, any of the depictions Slevogt created could hardly contrast more with the impeccably dressed Mohamed Nur one sees in the group photograph with Meinhof (plate 5) or as a teacher among the Somali show performers (plate 9).

In contrast to Slevogt's paintings or Nur's appearance on the postcard from the *Völkerschau*, his acoustic recordings, together with the texts that appear in von Tiling's publications, allow us to situate Mohamed Nur as a politically active person in the colonial history of the early twentieth century, and not merely a faint figure of imperial fantasy. Furthermore, much as is the case with Shire Rooble's recording, Nur's citation of Somali verses speaks of his reasons for leaving his country.

The Dervish movement disintegrated in 1920, after the British Royal Air Force bombarded their capital and headquarters in Taleh. Mohamed 'Abdille Hassan died in the same year. Shortly thereafter, in winter 1921–22, Mohamed Nur left Hamburg. He disappears from the German archives with this departure. One of the example sentences from *Somali Texte* (von Tiling 1918/19, 144) reads:

Im nächsten Winter will ich reisen (zur See).

In the coming winter I will travel (at sea).

Fragment IV

Josef Ntwanumbi: "We are initiates"
TRANSLATED FROM ISIXHOSA BY PHINDEZWA MNYAKA

O, haya we bawo, oh hayaye
O, haye bawo, o hayaye
Nako ndalamba
O, haye bawo wee
O, haye bawo wee
Nako ndalamba
Abakhwetha iyabawo wee

Abakhwetha iyabawo wee
Nako ndalamba
Iyo hayo hayo bawo wee
Hey hayo kub' haho he he he
Nako ndalamba
Abakhwetha iyo wee hee
Abakhwetha ndibo' hee hee
Abakhwetha ho, ndalamba
Salamba bawo bawo weehe
Wowu ha ba yibe ho weee hee
Nako ndalamba
Salamba bawo weee hee
Salamba bawo weee hee
Singabakhwetha
O, ndalamba
Salamba bawo weee hee
Salamba he bawo wee hee
Nanko nabakhwetha
Oh haye, hayi, singabakhwetha.
Oh haye, hayi, singabakhwetha.
Singabakhwetha, ngabakhwetha, oho hoo
Oko ndalamba
Inga' hoo, yaye
Nako ndalamba
Singabakhwetha hoho ha
Singabakhwetha oho hehe
Nako ndalamba

O, haya we Father, *oh hayaye*
O, haye Father, *o hayaye*
There I go hungry
O, haye Father *wee*
O, haye Father *wee*
There I go hungry
Initiates, Father *wee*
Initiates, Father *wee*
There I go hungry
Iyo hayo Father *wee*
Hey hayo it's bad *haho he he he*

2.3 Card for "Joseff Twanumbee" (Josef Ntwanumbi) from the register of internees at Ruhleben. Courtesy of the Red Cross.

> There I go hungry
> Initiates, *iyo wee hee*
> Initiates, *ndibo' hee hee*
> Initiates, there I go hungry
> We go hungry, Father *wee*
> *Wowu ha ba yibe ho weee hee*
> There I go hungry
> There we go hungry, Father *wee*
> There we go hungry, Father *wee*
> We are initiates, oh I'm so hungry
> There we go hungry *weee hee*
> There we go hungry *wee hee*
> There are also initiates
> Oh, no no, we are initiates
> Oh, no no, we are initiates
> We are initiates, are initiates, *oho hoo*
> I've been so hungry
>
> (Berlin Lautarchiv, PK 866/1)

Josef Ntwanumbi ("Twanumbee" in the Lautarchiv records) did not fight in World War I. He got caught by its outbreak when the ship on which he worked as a stoker reached the harbor of Hamburg. He was interned as a British subject in Ruhleben. Josef Ntwanumbi probably lived in barrack 13, the barrack that was assigned to the black internees in the camp (see plate 4).

The records of the Berlin Lautarchiv document that he came from the Eastern Cape in South Africa. Joseph Ntwanumbi spoke isiXhosa, English, and Dutch or Afrikaans, and was thirty-four or thirty-eight years old in May 1917. He had lived in India since 1897.

On the recordings, he sang of getting on the train, of hunger, uncertainty, and deprivation, a situation he describes as comparable to the situation of initiates (*abakwetha*). Neither photographs nor any other information apart from the identification cards issued by the Red Cross (see figure 2.3) have been found.

Albert Kudjabo and Stephan Bischoff

MYSTERIOUS SOUNDS, OPAQUE LANGUAGES,

AND OTHERWORLDLY VOICES

> Il faut combattre partout la transparence.
> (One needs to fight transparency everywhere.)
> —ÉDOUARD GLISSANT, *Le discours antillais*

Albert Kudjabo and Stephan Bischoff probably never met. Albert Kudjabo had fought on behalf of Belgium with the Corps des volontaires congolais until he was interned in the POW camp in Soltau, Germany, from whence he was transferred to Münster.[1] Stephan Bischoff had come to Germany to train as a dental technician and was living in Berlin when World War I began. Because he had come from the British Gold Coast of Africa, he was interned as a British subject in Ruhleben in 1915.

In the camps, Kudjabo and Bischoff were selected by Carl Meinhof to become "informants." Both Kudjabo and Bischoff had attended mission schools, and both complied with the requests of the linguist to a certain extent: they each knew and agreed to perform an encoded or "secret" language. Kudjabo performed a drum language from the Ituri region in what was then the Belgian Congo, and Bischoff apparently knew a spirit language of the Yeve society, Yevegbe, that was used in regions that are now part of Ghana, Togo, and Benin. Their acoustic echoes meet in the Berlin Lautarchiv. The avid interest of German linguists in languages they conceived of as *Geheimsprachen* (secret languages) points to their will to know what had so far eluded translation. On a practical level, the recordings with Bischoff and

Kudjabo speak both of the evangelizing mission and of the will to understand messages of long-distance communication in Africa and the Pacific islands.

The Martiniquan writer and theorist Édouard Glissant wrote that opacity resists demands for total transparency, which is a precondition of colonial control. Opacity, in this sense, rejects appropriation, observation, and control (Glissant 2010, 206). In her discussion of Glissant's concept of opacity versus transparency, Cecilia Britton (1999) turns to the etymology of the French term *comprendre* (to understand), indicating its relation to taking (*prendre*) as an act of appropriation. What the Königlich-Preußische Phonographische Kommission (KPPK) was able to collect in cooperation with "informants" in the camps—acoustic examples of encoded languages that were recorded and archived—was not as useful for their research as they had anticipated. Neither the recorded examples of drum language nor those of Yevegbe gave away the secrets of these opaque forms of communication.

Yevegbe is a spirit language that carries features of secrecy and exclusion, distinguishing initiated members of the Yeve society from the uninitiated. Drum language, which Albert Kudjabo performed on ten recordings, is based on the tonal language Bira and is primarily a form of long-distance communication that was used before the advent of Morse code or telephony (Storch 2011; Finnegan 2012). Drum languages used in African countries and elsewhere were perceived as secret languages by European travelers and linguists, who were unable to decode them. The drum language recorded with Kudjabo formed a barrier to imperial control of territories; Yevegbe hampered the conversion of people by colonial missionaries.

Neither coded language nor sound signals were translated for this chapter. Instead, I follow Glissant's (2010, 190) request "to focus on the texture of the weave and not on the nature of its components." In this case, the "weave" only partially refers to the functioning and structure of these languages themselves. Again, I follow the logics of the archive and of colonial knowledge production to make sense of the ways in which knowledge was generated and organized, the manner in which linguists claimed and obscured expertise, and particularly how utterances that addressed colonial violence came to be buried under colonial linguists' interest in secrecy.

Not all the recordings connected to Albert Kudjabo and Stephan Bischoff were performed as drum language or uttered in Yevegbe. Specifically, recordings that hinted at colonial violence in relation to the extraction of the riches of Ituri and to the project of evangelization in Ghana were reduced to "bycatch" by the linguists in their zeal to trawl for and identify specimens of secret languages. As with many other recordings from the Berlin Lautarchiv, critical comments were not registered in the documentation or in later publications, but they did survive as acoustic fragments.

Sounds African

> to take back all my names, shards of the past
> to remain the child of the mine and railroad
> family memory coupling with the locomotive
> exile in the bud, loneliness without end
> —FISTON MWANZA MUJILA, "Kasala for Myself"

3.1 Albert Kudjabo, photograph from family album (detail). Courtesy of Odette Kudjabo.

Albert Kudjabo drummed in a German prison camp on a slit drum from the Pacific. Wilhelm Doegen recorded his drumming, speaking, singing, and whistling in Münster, ten days after Kudjabo's twenty-first birthday, on March 25, 1917. A photograph, probably taken on the day of the recording, shows Kudjabo wearing a uniform and a scarf, bent over the drum that was borrowed from the Museum für Völkerkunde und Vorgeschichte (Museum of Ethnology and Prehistory) in Hamburg (see plate 11). The instrument appears to lie on frozen ground, probably in the courtyard of the camp.[2]

At the time the photograph was taken, Kudjabo had been imprisoned for almost three years. An extraordinary number of recordings in the Berlin Lautarchiv, thirty in total, are attributed to his name, apparently due to his skill and readiness to perform drum language for the linguists. Unlike many other recordings with African POWs, Kudjabo's recordings were actually revisited by the researchers after the war: his drum recordings were played on German radio in 1924; and Carl Meinhof published his findings on Bira (or Kibira, a language of the Congo region) in two articles in the 1930s (Meinhof 1938, 1939).[3]

The files in the Lautarchiv state that Albert Kudjabo was born on March 15, 1896, in a place Doegen listed under the name of "Dodo," in the Ituri region, which was then part of the Congo Free State under the rule of King Leopold II of Belgium (see figure 3.2).[4] In the documentation for the recordings, Doegen remarks that Kudjabo spoke and wrote Kibera, Kiswahili, French, and English. The files state that in 1905, at the age of nine, Kudjabo had moved to Kilo to attend the mission school there. The date of school enrollment connects Kudjabo's biography to the early days of a long history of violent economic extraction, which included the appropriation of farmland, forced removals of inhabitants, forced labor in mining and road construction, and the maiming and killing of millions of people in the Congo (Hochschild 1999). The mission school in Kilo was connected to the gold mine and run by the Catholic White Fathers. Kudjabo moved to Kilo in the same year that the gold mine there opened and the large-scale extraction of minerals that would transform the region irreversibly was initiated (see plate 12). When questioned about his profession before he became a soldier in World War I, Kudjabo's answer seems to outline the historical transformation of the region as he describes his movement from farming to "learning to handle mining equipment" ("Landwirt, beginnt zu arbeiten mit Bergwerkmaschinen"; PK 796). After his demobilization in December 1919, Kudjabo moved to the Schaerbeck district of Brussels, where he had to register with the Belgian Ministére de la Justice. He was subsequently placed under government surveillance.

According to the files of the Ministry of Justice in Brussels, Albert Kudjabo arrived in the country on August 4, 1914. He must have joined the Corps des volontaires congolais immediately.[5] The historian Griet Brosens (2013, 11) writes that Kudjabo suffered a head injury and was taken captive on August 23, 1914. The municipal record of Brussels Schaerbeck for 1920 gives *vulcanisateur* as Kudjabo's profession, which connects him to another lucrative form of colonial exploitation in the Congo: the infamous rubber industries.[6] Neither the archives in Belgium nor the files of the Lautarchiv account for the circumstances of his migration. How Kudjabo came to travel to Ghent at the age of seventeen remains unknown. One biographical date that was registered and archived coincides with the colonial history of Ituri. The Brussels files state that Kudjabo's father died in 1913. In that year, the colonial state clamped down on a local rebellion against deportations to and forced labor in the Kilo gold mines. The Belgian Army killed 291 people during this operation (Marchal 2003). Whether Kudjabo's father's death was related to this rebellion remains unknown. Yet one of the songs that was recorded with Kudjabo, which was falsely documented as "Die Leute haben meinen Freund unschuldig geschlagen" (The people have beaten my innocent friend) (Berlin Lautarchiv, , PK 798/3), might relate to the 1913 incident:

Bá mbétí nábhămʉ nkpáká bɪ́nzɔnɪ́
Bámʉtákánɪ́ hăshiyó kakănɪ́,
Kábúbandí ganí pa
Kálɪ́ kɪ́mabɔ́ pa.
Bámubétí kpaká bɪ́nzɔnɪ́.

Lfd. Nr.

PERSONAL=BOGEN

Lautliche Aufnahme Nr.: P.K.794   Ort: Münster
Datum: 25. 3. 1917
Zeitangabe: 2 Uhr 30 Min.

Dauer der Aufnahme: ____   Durchmesser der Platte: 27 cm
Raum der Aufnahme:
Art der Aufnahme (Sprechaufnahme, Gesangsaufnahme, Choraufnahme, Instrumentenaufnahme, Orchesteraufnahme): Trommelsprache
1.) Aufforderung zum Tanz durch den König 2.) dasselbe gesprochen.

Name (in der Muttersprache geschrieben):
Name (lateinisch geschrieben): K u d j a b o
Vorname: Albert   (K u d j a b o)
Wann geboren (oder ungefähres Alter)? 21 Jahre alt
Wo geboren (Heimat)?
Welche größere Stadt liegt in der Nähe des Geburtsortes?
Kanton — Kreis (Ujedz): Kilo, Kongostaat
Departement — Gouvernement (Gubernija) — Grafschaft (County): Dodo
Wo gelebt in den ersten 6 Jahren? bis zum 9. Jahre in der Heimat
Wo gelebt vom 7. bis 20. Lebensjahr? Dann zur Mission in Kilo, mit 17 Jahren nach Europa, Belgien, Gent
Was für Schulbildung? Missionsschule in Kilo
Wo die Schule besucht? in Kilo
Wo gelebt vom 20. Lebensjahr? Gefangen seit 1914
Aus welchem Ort (Ort und Kreis angeben) stammt der Vater? aus Gibeli bei Kilo
Aus welchem Ort (Ort und Kreis angeben) stammt die Mutter? aus Loko, Land der Babonya
Welchem Volksstamm angehörig? Babera Obersprache
Welche Sprache als Muttersprache? Kibera
Welche Sprachen spricht er außerdem? Suaheli, Französisch, etwas Englisch
Kann er lesen? ja   Welche Sprachen? Kibera, Suaheli, Französisch
Kann er schreiben? ja   Welche Sprachen?   "   "   "
Spielt er ein im Lager vorhandenes Instrument aus der Heimat? nein
Singt oder spielt er moderne europäische Musikweisen? ein wenig
Religion: katholisch Beruf: Landwirt, beginnt zu arbeiten mit Bergwerkmaschinen
Vorgeschlagen von: 1.
2.

Beschaffenheit der Stimme:
1. Urteil des Fachmannes (des Assistenten): Vortrefflich

2. Urteil des Kommissars:

3.2 Personal file of Albert Kudjabo. Berlin Lautarchiv.

> They have beaten my father for no reason
> They found him on a lonely path
> He has not provoked them with words
> He has not eaten their food.
> They have beaten my father for no reason.[7]

Read with the knowledge of the abject violence in the Congo, the song reverberates with bewilderment about an attack without provocation, without conceivable reason. In the archive this recording appears as a translation of drum language into Bira. In this case, the text should be read as a circumlocution: the tonal sequences of particular fixed phrases in spoken language are transferred into drum code (Finnegan 2012, 468). But what does it delineate? Unfair treatment? Senseless violence? Conflict? The title does not, as it does in other cases of translated drum language, provide a summary. This may be an error in the documentation of the Lautarchiv. Furthermore, Kudjabo speaks not of a friend, as indicated in the title, but of a father. The documentation of this recording also does not tell us whether or not this is a biographical snippet.

Carolyn Steedman (2000, 26) has addressed the historical incapacity of academics to hear genres in oral narratives, particularly those of working-class speakers, which were generally heard as biographical accounts instead. Ignoring the possibility of specific generic forms, researchers have often reduced spoken narratives to personal accounts sans literary form or aesthetic strategy. In a similar vein, many recordists of foreign languages (and music) knew nothing about genres in the languages they recorded and thus were unable to identify genres, not even in cases when they could understand the languages they recorded. Moreover, the chauvinism of Western researchers often precluded them from perceiving poetic language, specific literary genres, and metaphoric content (Hoffmann 2015b; 2020a). In other words, a spoken or sung text narrated in the first person does not necessarily indicate a biographical narrative. Elements of preexisting orature may have been recited in reference to past experiences that corresponded with the situation in the POW camps.

I have found very few academic publications on narrative genres in Bira. Most of the early ethnographic texts discuss the perceived physical otherness of the speakers. Information on Bira language and oral texts is rare and, where it does exist, it seems to be a by-product of anthropological research on foragers (Vansina 1990, 358).[8] Janusz Krzywicki (1984, 418) writes that orature in Bira has barely been researched. His more recent publication of a collection of narratives that were recorded during his research from 1975 to 1978 may be a reaction to this perceived lack of literature on the topic (2014). Krzywicki divides the narratives, which he recorded mostly with older men, into two genres, one of which is called *mákágáni*. This term might, as he mentions, refer to the situational context of storytelling. The term derives from the Bira verb *mákágán*, which roughly translates as "throwing news." According to Krzywicki, Bira narratives often engage with a disturbance of and the subsequent restoration of social order. All narratives in his compilation from 2014 are considerably longer than any of the recordings produced by Meinhof with Kudjabo.

Colonization certainly disrupted the existing social order in the Kilo area. Already during his childhood, Kudjabo had experienced the transformation of agricultural land into a huge alluvial mine (750 square kilometers) under military control. The large-scale disappropriation and displacement of people was followed by forced labor, writes Agayo Bakonzi (1982, 151). One of the results was a continuous food shortage because subsistence farming was no longer possible. Men from the area and from farther afield were compelled to work on the mines; women and older children were coerced into building roads to transport the exploits from the mines. The working conditions on the mines were brutal and included corporal punishment (174). Mine workers were chronically underpaid and suffered from sores on their feet and legs, without medical care. The mortality rate in the mines was very high, and laborers fled when they could (151). According to Jules Marchal (2003, 199), a system of forced labor, for which "chained laborers from all parts of the country were brought in," was already in place by 1907. Between 1905 and 1919, the Kilo-Moto mines extracted 23,434 kilograms of gold and generated more than 88 million francs for the colonial exchequer (126).

There were clearly many reasons to leave the gold fields of Kilo. Kudjabo managed to migrate to Belgium, yet he left the brutality of the colonial gold fields behind only to find himself on the battlefields of the Great War.

Kudjabo did not deliver a biographical account in the recordings made for the Lautarchiv. Yet the violence of colonial reign and the precarity of the loss of livelihood and migration are alluded to in several of his spoken and sung texts. These hints surface in two recordings as well as in the question, "Talala wapi?"—"Where will I sleep?"—which is woven into a chant that was recorded as a rowing song. As with Mohamed Nur's example sentence,

"In den nächsten Krieg wird niemand gehen wollen" (Nobody will want to join the next war), Albert Kudjabo's question glows like a firefly in the repetitive chanting of an improvised work song. The question ("Where will I sleep?") is inserted into the rhythm of a rowing song, which Kudjabo mostly hummed without words, and becomes part of the rhythm of traveling-for-work or traveling-as-work. We cannot know whether this sentence was the only fragment of the lyrics of a rowing song he could remember, or whether he inserted it into his version of a song as a comment on his present situation, in the way Josef Ntwanumbi inserted the phrase "Where am I dying now, Father?" into a song for the initiation of girls that was recorded with him in the camp (Hoffmann and Mnyaka 2015, 154). Kudjabo presented this rowing song as a chant that transmits a sense of repetition through rhythm, rather than by spelling it out textually. This small, reiterated textual fragment may refer to recurring experiences of displacement, upheaval, and terror in the young migrant's life.[9]

Kudjabo did not speak of the war on any of the recordings. War as a topic appears exclusively as a drummed code: *"Krieg" getrommelt* ("war" drummed), *dasselbe gesprochen* (the same spoken), *dasselbe gepfiffen* (the same whistled), recorded at 15:15 in the afternoon, on the day the photograph (plate 11) was taken (PK 796/1–4).

Ruth Finnegan (2012, 367 ff.) describes drum languages generally as systems of acoustic signs based on tonal languages, that is, languages in which the meaning of a word is distinguished by tone and not merely by phonetic elements (468). This means that in some cases, the same word or phonetic combination spoken in one tone or another may have entirely different meanings. Thus, to communicate clearly, drum languages cannot simply mime the intonation or sound sequence of spoken words. Because many words consist of an identical tonal combination, fixed modes of circumlocution are used to transmit meaning. Drum language does not abbreviate terms but rather extends them into fixed expressions, or phrases, to render them identifiable. Examples Finnegan (2012, 468) gives of these formulaic phrases in the drum language of Congolese Kele people are "manioc which remains in the fallow ground" for *manioc*; or "pieces of metal which arrange palavers" for *money*. Some of these fixed phrases bear similarities to proverbs. Others characterize sharply: *white man* is circumscribed as "red as copper, spirit from the forest" or "he enslaves the people, he enslaves the people who remain in the land" (468). In light of the long history of brutal exploitation and forced labor in the Belgian Congo, the latter description transmits a comment on colonial violence encapsulated within this fixed phrase. In the POW camp in Münster, Kudjabo first presented samples of drum language, and subsequently translated the sequences of sound signals he had played into Bira.

Drum Language, Indignation, and Colonial Control

Carl Meinhof's keenness to decode drum language surfaces in the transcription of an academic discussion on the topic that took place in Hamburg on May 11, 1916 (Thilenius, Meinhof, and Heinitz [1916] 1976),[10] about ten months before Albert Kudjabo was recorded by Doegen and Meinhof. By the time the discussion took place at a *phonetischer Abend* (phonetic evening) organized by the Phonetisches Laboratorium (Phonetic Laboratory) in Hamburg, Meinhof had already met Kudjabo, as is mentioned in a footnote to the transcription and, later, in Meinhof's correspondence with Felix von Luschan:

> In the prisoner's camp Soltau, near Hannover, I found a native from the Congo, who knows the drum language. I want to try to produce some recordings with him and for this I need three wooden drums of different sizes. Herr Professor Thilenius has kindly agreed to provide these drums and would probably be willing to go to Soltau[11] with me, to help me with the recordings. Because I would enter the field of ethnography with this, I would not want to do so without asking you for permission. It seems to me that in this language, like in Tschi, the pitch of the spoken language is inserted into the drumming, and this is the linguistic problem I am interested in. Of course, I will try to get as many drum recordings as possible with this opportunity. I therefore ask your kind permission to carry out this plan, or if you have any objections, to inform me of these. I am most keen to work, in this case, as usual, hand in hand with you.[12]

The transcription of the *phonetischer Abend* of May 1916 offers a glimpse of the debate around drum languages in German academic circles at the time.[13] Before an audience of fifty-five hand-picked guests, the linguist Carl Meinhof; director of the Museum für Völkerkunde und Vorgeschichte (Museum of Ethnology and Prehistory) in Hamburg, Georg Thilenius; and the musicologist Wilhelm Heinitz discussed their understanding of drum languages. Giulio Panconcelli-Calzia, the director of the Phonetic Laboratory, introduced the speakers. Heinitz (1941) would later publish on racialized features of African music and try to record drum languages with African POWs of World War II, who rejected his request. Thilenius had coordinated the Hamburger Südsee Expedition (expedition to the South Pacific) of 1908–10, which brought back

15,000 ethnographic objects for the museum,[14] among which was the drum that Kudjabo would later be given to play in the camp.

In this debate, two discursive elements stand out: First, drum languages are described by Carl Meinhof as requiring highly specific skills. Yet Meinhof's description of this technical mastery is followed by his immediate downplay of the complexity of this technique and the attribution of simplicity to this form of communication.

Second, drum language is characterized by its unnerving opacity, which was in fact disadvantageous not only to research expeditions but also to colonial military operations. These implications were considered particularly relevant. For Meinhof, research on African languages was clearly tied to practices of evangelization and colonization. In his earlier texts, he advocated for the benefits of language skills in the evangelizing mission and laid out these challenges as related to a poor command of the language by those who were to be converted (Meinhof 1905). Meinhof's ideas display a convoluted mix of research interests and practical concerns, which are juxtaposed with his patronizing assertions of the simplicity of the signals:

> I had the opportunity to get to know the drum language of the Duala in Kamerun. The result of my lengthy engagement with this object was the insight that the Duala are able to communicate a range of things by means of a large number of signals which are produced with wooden drums. One can get away with abusing an opponent from a distance in this way, and one is able to communicate any message with astounding speed, because the signals are passed on by other drummers. Protracted study is necessary for learning these techniques, and the common man usually understands only a few signals, for instance, the signal for "war," "come here," and his own name. Even dogs seem to be trained to come when their name is drummed. . . . On what principle the drum language of the Duala is based, I have been unable to determine as yet.[15]

Even after his long engagement with the Duala drum language, Meinhof admitted that he had as yet failed to work out exactly how the language operated and how it related to spoken language. The tension between the acknowledgment of complexity and skill and the notion of its simplicity is tangible in this passage. The dog motif reappears later in the debate: "Even a dog knows that. . . . One can easily train a dog to hearken to the drum call" (Thilenius, Meinhof, and Heinitz [1916] 1976, 1:26).

The ethnologist Paul Hambruch grudgingly reported that during the expedition to the South Pacific, drummed messages traveled faster than their steamboat. He called the expedition's inability to surprise the local population with their arrival a "mishap" (*Mißgeschick*), thereby connecting the strategies of the expedition implicitly with those of a military ambush. The topics of the speed of drum messages and the astounding detail conveyed in them are recurring themes in the discussion (Hambruch, as cited in Thilenius, Meinhof and Heinitz [1916] 1976, 1:27).[16]

Particularly, the opacity of this form of communication triggered suspicions regarding the content of messages that travelers could hear yet not understand. More than most other topics of linguistic research, the interest in drum language reveals a connection between colonial knowledge production and the military control of colonized countries.[17] Meinhof would later publish on Kibira (Bira), but not on the related drum language, which thus remained opaque to German linguists.

Wilhelm Doegen's interest in drum language relates to the air of the exotic attached to it. Recordings with Kudjabo were the first example of POW recordings he played during a German radio broadcast in 1924. The photograph of Kudjabo and the drum that had been brought to the camp from the Museum für Völkerkunde und Vorgeschichte (plate 11) was published in the *Berliner Illustrirte Wochenschau* in an article advertising Doegen's radio broadcast.[18] The caption reads "Kongoneger singt Ruderlied" (Congo Negro sings rowing song). Doegen describes Kudjabo's drumming as "hell—oder dunkel-dröhnend" (bright—or dark-booming; cited in Stoecker 2008, 125). While Doegen describes a booming sound, on the recordings I hear clunky drumming, which sounds much less resonant and more defined than the term "booming" would describe. The clear distinctness of tones is one of the preconditions for the reception and comprehension of a drum language based on tonal differentiation and connected to existing tonal languages. Doegen's boastful description of his recording of Kudjabo's drumming resounds with the enduring mystification and racialization of this form of long-distance communication, clearly expressing an intolerance of opacity, which was viewed as evasive. Doegen's choice of wording for the representation of the sound of the slit drum corresponds to Western imaginations of drum language depicted in colonial films,[19] but also in novels or comic strips, where African drum language appears as the ubiquitous signifier of the (aural) exotic and/or primitive, and as a sonic marker of blackness as lurking danger. Often in tandem with depictions of thick, impenetrable African forests, speaking drums were staged and apprehended as an acoustic sign of the opaque otherness of Africa. Drum language appears as a form of communication that literally could not be read by most white visitors and that baffled colonial explorers. The fact that any drum language would also be opaque to every African who did not understand this particular coded form of expression did not matter. In contrast to this, Kudjabo's equation of drum messages with telephone calls, which appears in the written documentation for recording PK 801/2 at the Lautarchiv, situates them firmly as contemporaneous with Western forms of long-distance communication.

During World War II, African prisoners in French camps angrily rejected the request by German musicologists that they demonstrate drum languages. Wilhelm Heinitz, one of the researchers in that German delegation, writes:

> Viele von ihnen nahmen die Zumutung, sich einer Trommelsprache bedienen zu können, fast als Kränkung auf. Sie sagten, dass sie als zivilisierte "Franzosen" doch moderne Nachrichtenmittel hätten und dass nur die "Heiden" noch trommelten und dass das Trommeln überhaupt nur eine Aufgabe mißachteter Berufe sei.[20]

> Many of them took the idea that they were able to perform drum language as an offense. They said that they, as civilized "Frenchmen," had modern communication devices at hand and that only heathens drummed, and that drumming was moreover assigned to men in professions of low esteem.

In German linguistics, drum language has had a long history of racialization and exoticization that would be worth a study in its own right; but that is beyond the scope of this chapter.[21]

Specimens and Spectacle

Albert Kudjabo's drumming was aired on German radio in 1924, seven years after he had been recorded playing the drum from the South Pacific.[22] The moderator of the *Funkstunde* (radio hour) introduced Wilhelm Doegen as "the professor who collected the voice recordings" and announced that Doegen would present some of the treasured recordings of the Berlin Lautarchiv. Doegen immediately interrupted the moderator, claiming that he had not merely collected, but actually created the recordings, which, he said, made a difference ("nicht er sammelt, sondern er schafft sie, das ist ein Unterschied!").[23] Rather perplexed, the moderator obediently repeated that Wilhelm Doegen had "*created* ancient languages, ancient types of music, rare kinds of music, words and sounds one would otherwise not get to hear."[24] The statement sounds absurd, but Doegen did not correct the moderator.

The philologist seemed to enjoy staging himself as the sole initiator and maker of the collection of recordings in the Lautarchiv, which was then called the Lautabteilung of the Preußische Staatsbibliothek (Sound Department of the Prussian State Library). As of 1924, the bulk of the acoustic collection of the Lautabteilung consisted of recordings with prisoners of World War I and some recordings with prominent people, including the German Kaiser and also Rabindranath Tagore. When the moderator asked Doegen to recall what had sparked the collection of the Lautabteilung, Doegen replied:

> Als nun die Kriegsgefangenen alle in Deutschland waren, da wurde auf meine Veranlassung eine große gelehrte Kommission ernannt, deren Präsident ich wurde. Dann wurden dort aufgenommen die Sprachen, die Musiken und die Laute von über 250 Völkern der Erde, im Bild, im Klang und im Text zugleich.

> When all the prisoners of war were in Germany, a large academic commission was appointed at my request. I became the president [of the commission]. The languages, musics, and sounds of more than 250 peoples of the world were recorded in image, sound, and text.[25]

As mentioned before, it was not Doegen but Carl Stumpf, one of the founders of the Phonogramm-Archiv in Berlin, who had acted as head of the KPPK.[26] Yet, without Doegen's initiative, none of these recordings, and thus none of these particular sentences, words, or sounds, would exist as acoustically archived items.

Before playing a recording of drum language, which Doegen considered to be an example of the most exotic and spectacular acoustic pieces in the collection, the radio broadcast touched upon several crucial points around the history of what is now part of the collection of the Lautarchiv of the Humboldt University in Berlin.[27] Doegen presented his version of the beginnings of the KPPK as a response to an opportunity—that of the presence of thousands of captured soldiers of the Triple Entente in German camps. His account speaks of the productivity and violence of colonial epistemological practices, which exceeded the moments of recording in the POW camps. Doegen's representation of the recordings, which does not identify the performers, is part of a systematic erasure, an appropriation: his account eclipses the men on whose expertise the recordings are based.

Neither the introduction of the sound files, played in the broadcast, nor the announcement of the program in the weekly paper *Berliner Illustrirte Wochenschau* on March 2, 1924, named Albert Kudjabo as a performer.[28] Here, as in so many choreographies of colonial knowledge production, a person of color's proper name was replaced with what I understand to be an ethnographic stage name, a marker that signified a racialized informant, removing "the faintest trace of the sovereign person" (Spivak 1999, 232). In many cases these stage names cannot be replaced by a proper name retroactively, because the person's name was either effaced beyond recognition by misspelling or not written down at all and thus absent from the archive. This iterative, often irreversible obliteration of names is an element of what Laura Stoler, via Derek Walcott, has called the "rot that remains" (Stoler 2008).[29] This absenting of names was violent. It was not a singular mistake or slippage but a regular practice of colonial knowledge production. As a systematic omission, it is part of the tenacious debris of the colonial archive.

In the broadcast, Kudjabo's drumming was presented as acoustic spectacle, a sonic *Völkerschau* moment with the promise of "spectacular and strange sounds" for the audience of the very new medium of radio. In this way, the reference to the recording of specific types of acoustically encoded messages in the plural ("sounds") conveniently linked the recordings as acoustic objects to racialized and phantasmagorical cultural origins. These origins with no specific cultural background were then tied to the fabrication of indexical blackness.[30] With the above reference to drum language, Doegen omitted Kudjabo's specific authorship along with his personal knowledge of a particular form of communication—the drum language Meinhof was interested in—in favor of recorded sounds as specimens and spectacle. This suggests that, apart from their spectacularity, recordings of drum language were regarded as desired objects of colonial knowledge production.

The way these recordings were presented for radio removed any trace of the performers' personal expertise. The specific knowledge of this language system, based on a particular tonal language (Bira, in this case) and the ability to perform it, were disassociated from the recordings as sound objects. In this way, Doegen, instead of Kudjabo, was presented on German radio as the expert on a Congolese drum language. This separation of the recordings from their performers flatly denied the cooperative, intersubjective character of knowledge production that took place in the camps.[31]

Picking apart these aspects of colonial knowledge production—namely, the dissociation of expertise from the acoustic object, the absence of Kudjabo's name as the creator of the recordings from the broadcast, and Doegen's denial of the intersubjective character of knowledge production—allows us to understand the process of distortion between camp, archive, and radio. It also offers insights into the constellation of the productivity of the commission's will to know and to conserve. Doegen's pavonine claim to have *created* this acoustic archive is at once true and false: the acoustic files, their composition and distortion, but also the accumulation of certain themes and genres, were indeed (also) a result of the productivity of a specific epistemic constellation of researchers and their interests, of funding, and of the "opportunity" of war and the internment of thousands of men. Yet Doegen's own contribution to the sound object he created with Kudjabo was limited to the practice of recording. The double inscription of sound files both as having been created for a European archive but also as carrying voiced fragments of archives of orature adds to the complexity of notions of "origins" and authorship with regard to these acoustic documents.

By means of some circular moves in epistemic practices and forms of staging—on the radio, in Doegen's publications, and in the weekly paper, and as well as in Meinhof's publications that represented him in public—Kudjabo was depicted as a "native," or, as Meinhof had it, as an "educated native" (Meinhof 1938, 147). In what I read as the autopoesis of the work of the KPPK, Kudjabo was equipped with a Melanesian drum, then unnamed, given a stage name, and stripped of his expertise. He was thus made to respond to the desires of imperial knowledge production, and his archival persona was then shaped according to the projections of German linguists.

Sotto voce: Sweet Potatoes and Gender Relations

Albert Kudjabo sang or spoke on few recordings. Obtaining translations for these was more difficult than with any of the other African language recordings I present in this book. This is because Bira is a language spoken only by a small group of people, both in the Democratic Republic of Congo (DRC) and in Uganda. After several unsuccessful attempts to find a speaker of this language in Cape Town, I was referred to the Dutch linguist Constance Kutsch Lojenga. She kindly agreed to take the recordings to northeastern DRC to have them translated. Unfortunately, in this case, I was not able to interact with the translators in person. When I met her in the summer of 2018, Constance Kutsch Lojenga told me that the two translators, Gilbert Katanabo Muhito and Faustin Sambu Avetsu, had struggled to make sense of the sound files. They had been unable to translate some of the songs at all.

Apart from the song most probably falsely documented as a translation of drum language (PK 789/3), there is a staged dialogue (PK 803/2). This recording was listed as "Geschichte von den Kartoffeln" (Story of the potatoes), here transcribed and translated by Gilbert Katanabo Muhito and Faustin Sambu Avetsu:

Nkálɨ́ amuulíshí mɨlɨ́ . . .
amuuli mɨlɨ́kɨ́makɨ́, atɨ:
-ngo kĕta a káu?
Bánɨ mɨlɨ́kɨ́ aeeee . . .
Bánɨ, bánɨ,
S nkálɨ́atɨ
Ingă.
kĕta ashíshie.
Bánɨmɨlɨ́kɨ́atɨ-íngă, kĕta káshí mbɛpa.

A wife asks her husband:
"Are there any sweet potatoes left?"
The husband replies, "Yes, plenty!"
The wife says, "No, they actually are finished."

The husband states that the sweet potatoes are not finished; he says he ate some yesterday and there were still plenty. They were not all gone. To which the wife replies that she did not eat them, but he can go and eat if he can still find any.

On first reading, this staged dialogue appears banal: a husband and wife bicker about whether there are, or are not, any sweet potatoes left. I was disappointed with the translation when Constance Kutsch Lojenga sent it to me. The German title given in the files of the Lautarchiv, "Geschichte von den Kartoffeln," had me hoping for a story about Germans or POW camps in Germany. Only upon reading Agoyo Bakonzi's 1982 dissertation did I understand that the massive shift brought about by the gold mine—which had taken away the arable land from Bira agriculturalists and had resulted in a crisis, particularly for the livelihoods of women—was being discussed in this recorded fragment, which may have been part of a longer narrative. Bakonzi writes that most Bira speakers were agriculturalists before mining began in Kilo. At the beginning of the twentieth century, the area was densely populated, and the main staple crop, sweet potatoes, was cultivated by women. The discovery of large deposits of alluvial gold in Kilo in 1905, and the drive of the colonial regime under Leopold II to extract the Congo's riches, meant that a large area of fertile land was transformed into a mine. Gold was skimmed from the topsoil, which destroyed agriculturally valuable land in the fertile area of Ituri. The dramatic shift in gender relations that accompanied the area's transformation from an economy based on agriculture to one centered on mining is not mentioned, not even implicitly, in most texts.

Yet the destruction of subsistence farming meant that female farmers became entirely dependent on their husbands' willingness to share what they had earned in cash and kind (Bakonzi 1984, 87). Colonization thus forced a European model of the nuclear family and of a shared household income, earned by the male head of a household, onto the people who inhabited the area or who had been forcibly relocated to the area to work on the mines. It was therefore not merely the patterns of land use shared by foragers, agriculturalists, and pastoralists that were disturbed but also the gender relations and familial structures of agriculturalists. The short scene of bickering reenacted by Kudjabo for the recording focuses on that recent, dramatic shift and its impact on a way of life that was destroyed by extractive colonialism. Comparable to his narrative of the man who was beaten for no reason, this echo of colonial violence is wrapped in a narrative that the linguists did not bother to attend to. In these two recordings, as with some of the recordings of Mohamed Nur as well as those of Stephan Bischoff (discussed below in this chapter), echoes of colonial violence entered the Lautarchiv as dispersed, isolated acoustic fragments. With Kudjabo, these concise narratives migrated from Ituri. The recorded fragments at the Lautarchiv may represent his only existing reminiscences of a life he left for good when he traveled to Belgium. The scene does not give an account of the history of Ituri. As with other forms of orature, it appears to pick out a crucial aspect of a historical or recent development and present it with no instruction on how to read or hear it. Yet if this is a fragment of orature, no explanation would have been needed in the specific settings of storytelling, which theorize politics and history. It is likely that a performative staging of a scene of domestic conflict would have been a form of processing the violent transformation of the region. The usual context where it would have been recited would have been significantly different from that of a recording situation in a German prison camp. Instead of the suspension of communication, an interested audience would have listened attentively. Perhaps other pieces of orature would have been recited or sung; performed texts would have engaged with others. While colonial linguists were keen on decoding secret languages, they did not expect to encounter modes of theorizing in oral texts and embedded in languages that were not written (Hoffmann 2012; Storch 2018). As a result, Kudjabo's texts did not speak to German linguists.

Back in Belgium, in 1919, Kudjabo cofounded the Union congolaise with the Congolese intellectual and activist Paul Panda Farnana and Joseph Adipanga, who had also fought in World War I. The Union congolaise, with Farnana as a chairman, was designed to support Congolese people in Belgium and in the Congo. Its members protested vehemently against forced labor and the ill-treatment of prisoners and called for the advancement of education for all Congolese people.[32]

Albert Kudjabo stayed in Belgium and married Ludovica Kempeneer in December 1925. They had four sons together. He died of a lung disease in 1938. His granddaughter Odette Kudjabo said he never spoke of the Congo.[33]

Sounds Unknown

> Western knowledge really does alter what it knows ... while also embracing the possibility that what resists such power, both from within and without, will, if given the room to speak, tell us something "we" are in no position to hear. About this, of course, we can *know* very little. Nevertheless, we must still do everything in our power to listen. —JOHN MOWITT, *Text: The Genealogy of an Antidisciplinary Object*

Stephan Bischoff commented on the evangelizing mission during a recording session in Ruhleben on April 27, 1917. His trenchant critique came wrapped in a fable that he presented as a language example to the linguist and former Pomeranian pastor Carl Meinhof. Into the concise, formal rigor of a fable, which illustrates and delivers a moral lesson, Bischoff inserted an actual place name, "Krachi." The name of this place connects his criticism of the work of Christian missions in his country to (then) recent events in German Togoland, specifically in Kete-Krachi, a regional center on Lake Volta, now in Ghana.[34] The recording entered the register of the Lautarchiv, with no indication of this critique, under the title "Die Geschichte von den 5,000 Affen" (The story of the 5,000 monkeys; PK 862/2). A century later, Johannes Ossey's attention to the recording and his 2018 translation disclosed a spoken text that engages with a series of violent events that connect imperialism and proselytization in a German colony. Bischoff's critique may have been prompted by Meinhof's interests: the recording produced on April 27, 1917, reverberates with Meinhof's concerns about the translation and translatability of Christian concepts and terms into African languages, which the linguist had spelled out in his book *Die Christanisierung der Sprachen Afrikas* (The Christianization of

3.3 (*opposite*) Signature of Stephan Bischoff from his letter to Carl Meinhof, August 23, 1918 (detail of figure 3.6). Personalakte Stephan Bischoff, 361-6-V-158, Staatsarchiv Hamburg.

3.4 (*above*) Personal file of Stephan Bischoff. Berlin Lautarchiv.

African languages) of 1905.³⁵ Meinhof used the recording session with Bischoff, who came from the town of Keta, to clarify certain terms that had created confusion and misunderstanding in the work of the mission.

Meinhof would come back to this topic when recording with Josef Ntwanumbi,³⁶ some weeks later, on May 19, 1917 (see Fragment IV). For these recordings, in isiXhosa, Ntwanumbi, the seaman from the Eastern Cape, counted, sang, chanted, and spoke. He also presented words that Meinhof had probably identified as useful terms with which to name the Christian god in this language (*Thixo, uThixo . . . inkosi, inkosin*) (Hoffmann and Mnyaka 2015).

On April 27, the conversation between Meinhof and Bischoff set the tone for further recordings. In the documentation of the Lautarchiv, Bischoff's fable is listed as his second recording on that day, coming directly after the recording of a series of words that related to Meinhof's interests (see figure 3.5). On the first recording, registered as "Synonyma" (PK 862/1), Bischoff pronounced words with similar-sounding vowels in his language, Ewe, and German (here in German and English):

Fünf (five)
Drei (three)
Dreinundfünfzig (fifty-three)
Fünfunddreissig (thirty-five)
Fünfundfünzig (fifty-five)
Blut (blood)
Knochen (bone/s)
Trommel (drum)
See (lake)
Weiß (white)
Töten (to kill)
Zerreissen (to tear apart)

On the recording one can hear Bischoff almost shout these words, first in Ewe, then in German. The performativity of his voice conveys an uncanny eagerness—perhaps the attempt to answer to the expectations of the linguist, who would request his release from the camp some weeks later. The absence from the archive of the request that prompted his speaking could open this sequence of words to speculation, and it may thus be tempting to read these words in relation to the violence of the war.³⁷ Yet Meinhof's book *Die Christianisierung der Sprachen Afrikas* (1905) forecloses such speculations. It

I.P.K. 862

Einzelne Wörter mit Tonunterschieden (- und hoch)

drei etɔ̄
fünf atɔ̈
33 blátɔ̄ vɔ́ ătɔ̄   +
35 blátɔ̄ vɔ́ atɔ̈
5. blátɔ̄ vɔ́ ĕtɔ̄   +
55 blátɔ̄ vɔ́ atɔ̈

Blut vu
Knochen ƒ u
Trommel vu
See ʃu
— fu
wuwu (das) Töten
vuvu (das) Zerreissen

3.5 Undated transcription attached to the personal file of Stephan Bischoff. Berlin Lautarchiv.

speaks of the linguist's motivation for recording these words pronounced by Bischoff: they sounded very similar, at least to the ears of the German missionaries and linguists, and had thus been a source of error and miscommunication during services and in sermons. Meinhof writes of the difficulty of hearing, distinguishing, and pronouncing intonation and pitch of words in Ewe:

> In Ewe ist ha "Schwein" und ha "Gemeinde" nur durch die Tonhöhe zu unterscheiden. Ein havi kann abgesehen vom Ton ein "Ferkel" oder ein "Gemeindeglied" sein. Das ist so verdrießlich, dass man für letzteres andere Ausdrücke gesucht hat. . . . Wenn der Eweküster das Lied ankündigt, dann wird auch ein geübtes europäisches Ohr manches Mal im Zweifel sein, ob es 33, 35, 53 oder 55 war. (Meinhof 1905, 51)

> In Ewe [the terms] *ha* for "pig" and *ha* for "parish" are distinguished only by pitch. Disregarding the tone, *havi* can thus mean piglet or congregant. This is so aggravating that one has searched for other terms for the latter. . . . When the Ewe sexton has announced the hymn, even for a trained European ear it is often unclear whether it is 33, 35, 53 or 55.

These disaccords must have caused disturbances beyond dissonant singing. Miscommunication led to indignation on the part of the congregants, writes Meinhof, yet one can also imagine ridicule and, perhaps, subversive laughter in response to the missionaries' failure to communicate.[38]

Meinhof's notion of the same word, "disregarding the tone," sounds rather absurd with reference to a tonal language. This explicit division of meaning from sound demonstrates how, as Mladen Dolar (2006, 19) has argued, phonology has killed the voice by reducing it to the function of a carrier. In Meinhof's text from 1905, which was written some years before he had established the Phonetic Laboratory in Hamburg, he described meaning as anchored in written signs or writable syllables. These he presented as stable, regardless of their intonation. Speaking was neither opera nor drum language, and meaning, despite knowledge that pointed to the contrary, could apparently not be imagined to reside in intonation. Tone, as that which voice produces, and which is intrinsically connected to the word and thus determines meaning, was reduced to a supplement, and merely marked (if at all) with an accent. In this way, piglets and congregation members were brought into aggravating (*verdrießliche*) proximity.

The demand for faithful translation of religious concepts and the ability to spread the Gospel with utmost clarity lay at the heart of Meinhof's treatise on the Christianization of African languages. Through language, the souls of potential believers in Africa were to be won. In this 1905 publication, Meinhof describes the mission's success as entirely dependent on competent translation for the successful transfer of concepts, an interpretation that does not account for the mission's imbrication with colonial violence. In his text, Meinhof ponders questions the historian Paul Landau critically addressed a century later: how exactly missionaries explained, translated, and transmitted their religious concepts, the Bible and the Gospels, and what exactly *was* understood by the recipients of such messages (Landau 2010). Aware of the pitfalls of translation, Meinhof demanded that sermons be held in clear and "natural" language (*natürliche Sprache*) and not delivered in *lingua sacra*. In his Hamburg lectures of 1911, which introduced the work of the recently founded Hamburgisches Kolonialinstitut (Hamburg Colonial Institute) to a wider public, Meinhof (1911, 95) presented "heathen cults" as speaking to their gods in oracle languages, in secret codes, and with feigned voices. In contrast, Christianity was to speak the language of revelation (*Sprache der Offenbarung*), without murmuring priests, secret languages, and other forms of secretiveness ("murmelnde Priester, Geheimsprache und Geheiniskrämerei"; 1905, 52).

The recordings with Josef Ntwanumbi and Stephan Bischoff show that, more than a decade later, Meinhof still pursued these questions actively. That the words Bischoff spoke into the phonograph correspond with the discussion of the very same terms in Meinhof's *Christianisierung der Sprachen Afrikas* also demonstrates that Meinhof was well prepared for the recording session on April 27, 1917. Perhaps he had planned to recruit new language assistants even before he met Bischoff and Mohamed Nur. By the time of Meinhof's visit to Ruhleben, the previous Ewe teacher at the Hamburgisches Kolonialinstitut, Victor Toso, had been deceased for a year.[39] Stephan Bischoff would become his successor at the end of 1917, but this might not have been decided at the time the recordings were produced.

There were also other recordings made with Bischoff. Recording number PK 863/1 conserves a series of formulaic salutations, which resound oddly in an internment camp:

Guten Morgen euch allen!
Guten Morgen
Sind alle wohlauf daheim?
Wie geht es den Kindern?
Es geht Ihnen gut!
Seid ihr wohlauf?

Wir sind wohlauf.
Das freut mich.

Good morning to all of you!
Good morning
Is everybody well at home?
How are the children?
They are well!
Are you well?
We are fine.
I am glad to hear this.[40]

On the same day as the list of words, one or two songs in a spirit language,[41] which the files describe as "Geheimsprache des Yefekultes" (Secret language of the Yeve cult), were also recorded. It is important to note that Bischoff did not refer to the language as a secret language. Instead, he spoke of *engaging with So*, a deity of the Yeve pantheon (PK 863/3). Johannes Ossey translates his announcement as follows: "I will sing to you a song in Yeve. It engages with the deity named So." Details of whether or not he was asked to say more about this recording, whether Meinhof intended to decode it, and how exactly Bischoff came to present it did not enter the archival record.

Sacred to Secret

Colonial linguistics did not respect the boundaries of spirit languages, nor did German linguists describe these as spirit languages. Instead, writes Anne Storch (2018, 105), there was a particular eagerness to prize open every secret of the languages in question and leave no mystery intact. Carl Meinhof had certainly read Jakob Spieth's publication on the religion of Ewe speakers (*Die Religion der Eweer in Süd-Togo* [1911]). Spieth's extensive chapter on Yeve secret societies was apparently the most recent German text available on this topic in 1917. It is mostly based on a report by the Togolese evangelist Hiob Kwadzo Afelevo, who had been initiated into the Yeve society before he converted to Christianity.[42] Spieth, who was a missionary of the Norddeutsche Missionsgesellschaft, described Yeve society as creating a space where magical practices, immorality, sexual transgression, and crime thrived. In his report, So appears as the god of thunder, hailing from a place called Hebie. So's emblem was an ax, which cracked houses, trees, and people alike (Spieth 1911, 173). Particularly with regard to Meinhof's recordings of the encoded language called Yevegbe, it is important to note that, per Spieth (1911, 184), the betrayal of the secrets of Yeve society was forbidden under penalty of death.

While Hiob Kwadzo Afelevo's father was a member of the Yeve society and Afelevo had himself abandoned Yeve to join the mission, Stephan Bischoff's father was an evangelist, and Stephan Bischoff had probably been a Christian before he was initiated into Yeve society. To my knowledge, no account by Bischoff about his own engagement with Yeve entered the archive. Yet, because he was able to lend his voice to So, to perform songs in Yevegbe, a language that was exclusive to members who had learned it in the process of initiation, one can assume that Bischoff must have been a follower at some point. Based on Afelevo's report, Spieth (1911, 183) writes that Yeve language(s) were unknown to most of the initiates when they entered the convent (*das Kloster*). According to the rules of the shrine, they had to converse exclusively in this language during the period of their initiation.

What does not surface in the archive is what, exactly, it may have meant for Bischoff to summon So with his drumming and to lend his voice to a language that was supposed to be strictly confined to conversations with spirits or deities in exclusive spaces connected to Yeve society. Did Bischoff consider it safe to call upon So and to allow this language to be recorded, even though Yeve society strictly prohibited any communication of its secret knowledge to the uninitiated? Did he trust in the opacity of the spirit language to shield sacred or secret content from the ears of linguists and fellow inmates? Was he coerced? Or did he feel compelled to offer this specific performance?

Bischoff's performance of Yevegbe certainly contributed to a set of abilities that must have made meeting him seem to be a tremendous stroke of luck for Carl Meinhof: according to the files of the Lautarchiv, Bischoff spoke Ewe, Ga (Gã), Adangbe (Dangme), English, and German. Bischoff's father had been a religious teacher with the Norddeutsche Missionsgesellschaft in German Togoland. Additionally, at least from Meinhof's point of view, Bischoff was able to speak the language of a religious society in which German linguists were extremely interested and that the Norddeutsche Missionsgesellschaft had tried to ban (Meyer 2002, 188).

Meinhof had presented his ideas on the role of language in colonial power relations a decade earlier at the Deutsche Kolonialkongress (German Colonial Congress) of 1906. In his address, he agitated against teaching German to a wider public in Africa because, in his view, this would shift the power relations in the colonies. Allowing "Natives" (*Eingeborene*) to understand the communication between German colonial masters (*Kolonialherren*) would also enable them to inform their countrymen on the aims of the Germans

(Esselborn 2018, 124). At the same time the "mother tongue" (*Muttersprache*)[43] of the colonized population would become a secret language that could evade the understanding of the colonizers. Thus, teaching German, Meinhof contended, would be a grave mistake, almost an invitation for rebellion. In the process of gaining control over colonized territories, all languages that were not (yet) understood appeared as a strategic disadvantage for the German military, for colonial administration, trade, and the evangelizing mission. Spirit languages in particular, which the linguists described as secret languages, were not taught to the uninitiated and could be used to subvert colonial control.

For the German missionary linguists, Yeve society was interesting in terms of its cultural, linguistic, and religious knowledge, which had probably migrated to Ghana and Togo from Dahomey in the nineteenth century during a time of increased imperial expansion in the region (Zimmermann 2015, 275; Spieth 1911, 177). Meera Venkatachalam (2012) writes that, from the 1850s onward, the popularity of Yeve grew in the area around Keta, where Bischoff was born in 1891. This means that in the 1850s two novel cults emerged simultaneously in the area: Yeve and Christianity (Meyer 2002, 180). Particularly around that time, the relatively new Yeve shrines were open to everyone, and membership in Yeve societies was not organized along lineages. The rise of Yeve, writes Venkatachalam (2012, 109), corresponded with the rise of a (by then) illicit, regional slave trade, and shrines were used to continue this trade. Yet Yeve society also accepted people who could not claim a lineage, and thus offered options of identification for formerly enslaved people.[44] While the recordings in Yevegbe indicate that Bischoff probably had been initiated at some point, it is uncertain what attracted him to Yeve and how this related to his Christian upbringing or his identification as Christian, which appears in the files of the Lautarchiv (see figure 3.4).

Bischoff may have been one of those people who could not claim affiliation with local lineages. The European names of his parents, Friedrich Bischoff and Marie Jonas, may be related to the practice of the Norddeutsche Missionsgesellschaft of buying children from slave traders in order to evangelize them.[45] Birgit Meyer writes that this was done because, particularly in its early years, the mission was unable to attract Ewe speakers willing to convert. Lacking followers, the missionaries devoted their funds and energy to the education of these purchased children who would go on to become mission assistants. The Missionsgesellschaft used this coercive practice of recruiting missionary assistants and congregation members for more than a decade, between 1857 and 1868 (Meyer 2002, 182).

In the files of the University of Hamburg (to which I will return), Meinhof describes Stephan Bischoff's father as a "native evangelist teacher." Because Bischoff was born in 1891 (or 1892) and because both his name and his parents' names sound German, it is possible that he may have been among the second generation of mission-educated children. His language skills in English enabled him to later work at the railway, which points to his education at the mission, where pupils were instructed in German and English (Meyer 2002, 182); by contrast, at the village schools, the missionaries instructed pupils in Ewe.

Venkatachalam (2012, 111) states that Yeve offered status to social rejects and people who were stigmatized. On a practical level, this means that Yeve created safe spaces for its followers, which allowed them to evade not only colonial but also parental control. From Spieth's (1911, 185) writing, I gather that, specifically for young women, Yeve offered options to escape arranged marriages; but it also likely allowed for a new start, perhaps even offered a new identity to all followers, whose past was ritually buried upon initiation (197). Venkatachalam (2012, 111) writes that Yeve communities also "functioned as business enterprises into the 1900s, and initiates were involved in the production and trade of copra, palm oil and liquor." Perhaps Yeve society offered an attractive option for Bischoff, who might not have had a socially acceptable lineage if his parents had, indeed, been enslaved as children purchased by the Norddeutsche Missionsgesellschaft.

Yeve society was neither ancient nor static. Nor did association with Yeve exclude being Christian. Instead, the two cults appear to have offered a set of choices that enabled followers to navigate the changing economic and social conditions, providing ways to respond to and reflect on experiences of the societal transformations of the time in the region (Venkatachalam 2012, 113). Together with its creation of exclusive spaces, the encoded language Yevegbe allowed the Yeve society to resist German colonization.[46]

While colonial linguists, following missionaries and colonists, heard a secret language (*Geheimsprache*), Anne Storch (2011, 89) describes Yevegbe, the language connected to Yeve, as a widely known lingual practice by means of which gods spoke (or speak) through a possessed person to communicate with their followers. Speaking Yevegbe performatively indicates an altered state of being as well as a modification of social relationships after initiation: it distinguishes the possessed, initiated person from the uninitiated, those who do not speak or hear (understand) the language of the spirits. Yevegbe, writes Storch (2011, 89), is based on Ewe (or Ewegbe). Storch argues that in the state of spirit possession, a person does not control language but becomes instead a medium for the spirit or deity who speaks through that person. Crucially, Storch attends to what Glissant has described as "the weave" of opacity. She does not merely describe the encryption that transforms Ewegbe into Yevegbe. In Storch's work, the deciphering of a secret code is not a gesture of victory over opacity (as a nuisance or a challenge). Rather than focusing on decoding as merely an act of translation, she unpicks the operation of codification as a moment of theorizing, in which a spoken language is intrinsically transformed with expressive elements in a movement toward a temporarily altered state of being in the world. She shows that the linguistic deviations that transform Ewegbe to Yevegbe carry meanings that extend the lexical significance of words. One example is the change of the use of the personal pronoun from "I" to "person." The transformation of the speaker to "person" signals the visit of the spirit, who now speaks through the possessed. The speaker thus becomes the medium for the spirit's voice (Storch 2011, 100). Other elements of this modification of speech include the use of Latin words and words from other languages spoken in the region. The insertion of these words signals the incorporation of alterity, which includes the otherness of imperial invaders and the language of their religious cults. The switch of active speaking positions—from a person who speaks to a deity to a spirit that speaks through the person—also entails prosodic alterations. Audible breathing or the sound of clearing the throat signal the arrival of the spirit who takes over the body of the speaker. On the recording, the content of which Bischoff describes as his engagement with So, Bischoff's voice sounds hoarse, raspy, breaking, and definitely different from the way it sounds in his other recordings. In terms of the epistemologies of spirit possession, this implies that it is not Bischoff one hears in the recording but rather So, who is speaking through him.

Western knowledge production alters what it sets out to know, as John Mowitt (1992) explains. Colonial linguists were certainly in no position to hear what did not belong to their field, nor what was impossible to perceive from their epistemological position as German linguistics in the early twentieth century. Because the possibility of embodying a speaking deity did not occur to the linguists, they heard Bischoff sing in a feigned voice. Yet, when asked for the Lautarchiv records which languages he spoke, Bischoff did not mention Yevegbe. He did not mention it because he did not actively speak this language. By means of his drumming he summoned the deity. On recording PK 863/3, one hears So, speaking through Bischoff.

As has been shown in this book, the content and forms of spoken and cited texts that were presented to the KPPK by POWs have been systematically overwritten by the classifications of the recordists. This durable distortion of content in the documentation of the Lautarchiv is a notable result of a practice of knowledge production and archiving; and while the content of many spoken texts has been erased from the register, in this case the coloniality of knowledge production omitted a speaking deity. A deity summoned by drumming, whose arrival is signaled by the subsequent shift of speakers, clearly explodes the logic of this archive but also confounds the rules of evidence. Here, the person who appears to speak into the funnel is not the speaker, and the speaker remains invisible. Visual evidence, which would normally stabilize the connection of voice to a speaking body, is not available in the case of a visiting spirit. Instead, a singing (or speaking) spirit appears as the absolute *acousmêtre* in this archive of acousmatic voices. Here, recent Western theorizing on voice reaches its limits: even before the practice of recording severed the human voice from the speaker, this voice was bodiless. Only the coarseness of this voice's materiality, its "grain," together with the language in which it communicates, signal the entrance of So, who possesses Bischoff's body and speaks through it (Barthes 1979). This does not merely undo persistent ideas of the authenticity of voice as a trace, nor does it delink voice as an acoustic phenomenon from voice as agency, the will, and the self. It also rejects the notion of voice as intrinsically connected to the human body that utters it.[47] At the time of this writing, it is not clear to me what this means with regard to listening to Bischoff's recordings. Is a German archive permitted to hold recordings of a West African deity? Am I allowed to hear this? Did Bischoff expect anyone to decode the recording? So far, I have found no sign of any decoding attempts either in the archive or in any of Meinhof's publications. Unquestionably, the opacity of the manipulated language needs to stay intact to screen So's speaking from Western academics' interpretations.

Already in 1894, long before he met Bischoff in Ruhleben in 1917, Meinhof had written about secret languages.[48] Like drum languages, spirit languages were perceived to be "secret languages." Both presented an impenetrable barrier for German linguists and also for missionaries and colonial administrations. Unlike drum languages, spirit languages distinguish initiated speakers from noninitiated people. In 1894, Meinhof described the available knowledge on secret languages as meager (*dürftig*). This, he wrote, was related to the situation where these opaque forms of communication were known only by a few natives (*wenige Eingeborene*) and were obscured by a veil of religious and magical darkness ("vom Schleier religiöser und zauberischer Dunkelheit umwoben"; Meinhof [1894] 1976, 1:151). It was therefore not easy for linguists to gain information on these languages. As with drum languages, Meinhof first affirmed the complexity of a code that "no European has mastered," only to subsequently mitigate this statement by claiming that the "surprising uniformity of African thought" ("die erstaunliche Gleichförmigkeit des afrikanischen Denkens") would allow for a swift decoding of secret languages (1:151). Although Meinhof set out to give an overview based on the knowledge of secret languages that had been gathered so far, describing encoded languages of various kinds that were used by sacred and secular secret societies and guilds, he seemed particularly interested in the secret languages of religious groups. These he described, referring to texts by African missionaries as deceitful and as used particularly for immoral ends or acts of revenge (1:152–55).

Toward the end of his short presentation on secret languages, Meinhof mentioned *Gaunersprache* (argot) in Germany, observing that some of its words had migrated into the *Volkssprache* (national language).[49] While writing, publishing, and standardization via dictionaries would protect the *Volkssprache* from such intruding words, oral languages were less protected. Oral languages, Meinhof argued, were porous and thus vulnerable, unable to avert the intrusion of words from encoded languages ([1894] 1976, 1:157). Meinhof's comments not only turned Ewe into a national language (like German was in Johann Gottfried Herder's sense; see Meyer 2002), but also discredited African secret languages, both religious and mundane, in line with his understanding of argot in Germany. The latter he described as the dubious codes of hustlers, traders, and mobile people in Germany, which he stigmatized and distinguished from his ideal of the *Volkssprache* as stable and localizable within the borders of the nation. Meinhof's characterization of secret languages in Germany transferred their disrepute onto what he understood to be comparable secret languages in Africa. His classification of codified languages was thus not interested in their function for their speakers but focused instead on the barrier they formed to being understood by uninitiated listeners. For Meinhof, the "intrusion" of argot into national languages, and thus into the soul of a people, was undesirable because it carried the stigma of marginalized groups, of itinerant people like the Jenisch in Germany, for example. In a similar vein, the migration of words from secret languages into the repertoire of what Meinhof understood to be national languages in Africa was presented as objectionable. In this way Meinhof transplanted the Romantic idea of national language and its unifying effect (for a nation-state) as well as the bad reputation of codified languages in Germany into his description of secret languages in Africa, vaguely addressing them as a threat to languages presumed to be a stable and definitive part of national identity.

From Ruhleben to Hamburg and to Prison Again

Most of what can be known about Stephan Bischoff surfaces in his personal files in the Lautarchiv and in the files of the Hamburgisches Institut für Kolonialsprachen, now held at the Staatsarchiv in Hamburg.[50] Like Albert Kudjabo and Mohamed Nur, he had migrated to Germany before the war began. How Bischoff traveled is not clear, yet he had already been living and working in Berlin when he was interned as a foreign subject in February 1915. The British list of inmates at Ruhleben registers his address as Auguststrasse 4 in Berlin, and marks him as "colored."[51] The documentation of the Lautarchiv (PK 862; see figure 3.4) states that he was born 1891 in Keta, in what was then the British Gold Coast (now Ghana), that he had attended mission school there and in Lomé, was of Lutheran Christian denomination, and spoke five languages. The Lautarchiv file gives his profession as *Lademeister* (responsible for cargo) whereas a British list of inmates of Ruhleben registers him as a porter (and gives a different date of birth).

The traces of Bischoff in German archives thicken with Meinhof's attempt to employ him as a language assistant at the Hamburgisches Institut für Kolonialsprachen. In early May 1917, less than three weeks after he had spoken into the phonograph in Ruhleben, Carl Meinhof wrote to the Oberschulbehörde (board of the university) asking for support in requesting Bischoff's release from the camp. I quote this letter at length because it testifies to the importance of "language assistants" not merely because of their teaching but also because of their work as informants and translators of correspondence with Ewe speakers. Speakers of African languages were thus an integral part of ongoing knowledge production on Germany colonies—in this case German Togoland—at the Hamburgisches Kolonialinstitut.

> I hereby would like to ask the board of the University to grant me permission to employ a language assistant for Ewe, to substitute for Victor Toso, who passed away. For 1.5 years I have had to provide Ewe lessons without the support of a language assistant, a situation which has caused the instruction to suffer considerably. Additionally, I have grievously missed the cooperation of a language assistant in the publication of Ewe texts, such as the first issue of the *Zeitschrift für Kolonialsprachen* [Journal for colonial languages], and for the translation of Ewe texts. So far my attempts to find a replacement have been unsuccessful, but now I have found a suitable candidate in Ruhleben. Stephan Bischoff was born in Keta, British Gold Coast, and speaks Ewe as his mother tongue. His father was an evangelist at the Bremer Mission and

Stephan Bischoff speaks German and appears to be an educated and sensible person. He worked on a steamboat and has been interned since the beginning of the war. I believe that he could be released or suspended, and therefore ask respectfully (*gehorsamst*) that a request for this be sent to the director of the British Camp (near Spandau). I propose to remunerate him with 130 marks monthly, as with the assistants Messi and Makembe, and to employ him for this semester. Several [female] teachers of the Norddeutsche Missionsgesellschaft have expressed interest [in Ewe classes], yet they plan to enroll only if an assistant is available, because they have been in Togo and already have basic knowledge of the language. Without an assistant I will not be able to provide conducive lessons for them. So far nobody is interested in a beginner's course in Ewe. Additionally, the cooperation of the language assistant will be of use in informing us of the situation in Togo and to translate incoming letters.[52]

The letter articulates the need for Ewe instruction in Hamburg and shows the institute's strong connection to the Norddeutsche Missionsgesellschaft, which had been active in German Togoland since 1847 (Meyer 2002, 172). Bischoff's ability to speak German corresponds with his mission education in Togo and the fact that he grew up in a mission-educated family. Bischoff's file at the University of Hamburg, generated in September 1917, gives the names of his parents as Friedrich Bischoff and Marie Jonas, states that he had worked for the railway in Togo, had been a cocoa inspector, and had worked in a shoe shop called Zebra in Berlin before he was interned in Ruhleben as a British subject. In the file he appears as a *Nichtwissenschaftlicher Angestellter* (nonacademic employee), which informs on the status his language skills and expertise had at the University of Hamburg.

After Meinhof's first request to release Mohamed Nur and Stephan Bischoff from Ruhleben was rejected in June 1917, a second request was granted in September 1917. The institute's contract with Bischoff established that he had to be present for four hours per week of grammatical instruction by Meinhof, and he also had to offer ten hours of conversation with students per week. In addition, he had to make himself available as a language informant ("hat als Sprachquelle zu dienen") and as an informant for the seminar of colonial law.[53] Upon his arrival at the University of Hamburg on September 24, 1917, Bischoff received an advance payment of 40 Reichsmark. In December 1917, the board of the university further requested that he and Nur be exempted from the duty of presenting themselves at the Fremdenpolizei (foreign police) twice weekly, because this would disturb the teaching schedule. The letter from the board further asks that the "two colored language assistants" ("die beiden farbigen Sprachgehilfen") who were on leave from Ruhleben be permitted to leave their houses after 8 o'clock in the evening to attend concerts. The request, signed by Meinhof, states that the Institut would take responsibility for the "unobjectionable behavior of the Negroes" ("das einwandfreie Verhalten der Neger"). While the status of Nur and Bischoff was juridically related to their position as "foreign subjects" in Germany during the war, this sentence marks them as men of color. The document also addressed concern about possible "objectionable"' behavior, which is presented in relation to their skin color.[54]

The remarks in the document testify to the systemic racism in Germany at the time, which corresponds to, but certainly was not entirely caused by, propaganda against African soldiers, particularly against those in the French army (Koller 2004; Frobenius 1916; Hoffmann 2014a). While Bischoff did not work in a *Völkerschau* or appear in exoticizing images, or in fact in any image I could find, in his archival trace he is repeatedly marked as a person of color, and he may have been employed at the Zebra shoe shop in Berlin because he was black.

A year into his employment in Hamburg, a much more intense discussion on Bischoff entered his personal files at the university, which, again, was informed by racist perceptions. On April 1, 1918, the Polizeibehörde (police department) of Hamburg informed the university that Bischoff had been in detention awaiting trial since March 15. He was accused of trading stolen goods, namely, clothes and haberdashery. Shortly thereafter (on June 22) he was sentenced to a year in prison. During the proceedings of the court case, the spotlight on Bischoff intensified. His personality, character, and biographical details were drawn into the light of official appraisal because it was assumed that he had trespassed against the law. Comparable to the men in Foucault's "Lives of Infamous Men," Bischoff was touched again, this time more sternly, by the "claw" of power (Foucault 1967). Accordingly, his archival trace thickens; his character, morality, and usefulness for the institute were discussed by the university's board, by the Hamburgisches Institut für Kolonialsprachen, and by the court of the city of Hamburg. These dialogues of official appraisal create and preserve the figure of a native informant from German Togoland. The files of the actual court case against him and his German codefendants no longer exist in the Staatsarchiv Hamburg.

Meinhof strategically petitioned the university's board for support in requesting a pardon for Bischoff. Meinhof argued that Bischoff did not know that the clothes he sold in Hamburg were stolen goods. He also contended that Bischoff was not capable of making sense of a German trial and characterized him as "a bit cumbersome but good natured" ("etwas schwerfällig aber gutmütig"). Read through Meinhof's letters, Bischoff figures as a "good native," whose intellectual incapacity got him into trouble. Meinhof's clichéd characterization stands in contrast to the fact that Bischoff had worked for the German railway in Togo for years and that he had come to Germany, probably on his own, with the aim of training as a dental technician. Apart from testifying to his character as an expert on "Africa," Meinhof professed Bischoff's indispensability. For this he brought to bear the situation that the Institut had to cancel the Ewe lessons, which positioned Bischoff as irreplaceable, or at least as a very useful "native informant." In response, the Oberschulbehörde deferred to the magistrate of the city of Hamburg in support of Meinhof's wishes and arguments, since "the professor is an expert on the psyche and the educational status of natives" ("der Professor ist ein Kenner der Psyche und des Bildungsgrades von Eingeborenen"). On October 4, 1918, the city's Senatskommission für Justizverwaltung (Senate Committee for Administration of Justice) informed the university that Bischoff would be released on parole until 1921. In November 1918, Bischoff returned to the Institut. He was reemployed yet no longer granted a permanent contract, working on a day-to-day basis. During the same month, Meinhof informed the board that he expected Bischoff to return to Togo in January 1919. Only one letter from prison, handwritten by Bischoff himself, unsettles the solid stack of letters and documents that discuss his character and that speak about him without his input. This letter appears as the most personal trace of his presence in Hamburg: The letter, on which a seal shows that it passed censorship, speaks of hunger and hope, and expresses wishes for himself and for Meinhof (see figures 3.3 and 3.6).

SENDER: STEPHAN BISCHOFF
ADDRESSEE: HERRN PROFESSOR MEINHOF
THE SENDING WAS PERMITTED BY THE JUDGE ON AUGUST 23, 1918.
[UNTERSUCHUNGSGEFÄNGNIS (REMAND PRISON) AUGUST 23, 1918]

Sehr geehrter Herr Professor,

Your dear letter of the 18th of August I received with joy. I am happy [to hear] about the animals in the forest. Could one not slaughter them all and send them here to Hamburg? We need meat to eat now. It is good that your son came back and we hope he will recover. With regard to my case, I still wait in good faith for the court [to decide]. I am not too well, we get very little to eat here, apart from this, I am fine. The overseers [*die Herren Aufseher*] treat us well. I hope these bad times come to an end soon. One cannot keep on living like this. No other news from my side. Regards to your wife.

Your obedient boy [*Ihr gehorsamster Junge*] Stephan

PS I really hope the Germans get their colonies back soon, then I will go home immediately, I miss my mother and brothers,

Ihr Bischoff[55]

5,000 Monkeys, or How to Serve Critique on the Evangelizing Mission to a Pomeranian Pastor

Stephan Bischoff's letter, written after Meinhof had obtained his release from Ruhleben and had engaged a lawyer to get him out of prison, reads as timid, with a subordinate tone, which may have been strategic. Although he was a prisoner in the moment of writing the letter and similarly had been an internee in Ruhleben when the acoustic recordings were produced there, the story of the 5,000 monkeys, in which he criticized the evangelizing mission in Togo and the conduct of the Germany colonial army in Krachi, comes from an entirely different position from that of the written letter. Clearly, his letter is a personal address to Meinhof, on whose benevolence Bischoff now depended more than ever. The speaking position of the fable differs considerably from his role as the subservient protégé.

Regarding the spoken text, Johannes Ossey and I have found it difficult to identify a genre. Yet, like the recordings with Josef Ntwanumbi, Bischoff's spoken text could be based on a specific genre that had been altered or adjusted to the situation of recording. The historian Wazi Apoh (2020, 464) writes that orature around the event to which this fable refers still circulates after more than hundred years. Thus, while Bischoff clearly is the speaker who presents the fable, he may not have been its author. Given the prominence of the events to which this fable refers, it is very likely that it was part of a swarm of utterances that made sense of outrageous events and a massive shift of power relations in the region around Krachi. While these events still feature in orature around Krachi, only a small fragment of what may have been a discursive formation in orature entered the Lautarchiv. Presenting the fable as an element of a local discourse and of orature may have invested

Absender: **Stephan Bischoff** Unters Gef. Hamburg
Empfänger: Herrn Professor Meinhof

Eingegangen
den 2 3. AUG. 1918
Landgericht H...
Strafkammer 1

Absendung richterlich genehmigt:

Notiz

Die Annahme von Wäsche findet nur Sonnabends von 2—5 Uhr nachmittags statt.

Untersuchungsgef 23. August 1918

Sehr Geerter Herr Professor

Ihr werter brief v. d. 18. August habe ich mit freude erhalten, es freud mich sehr über die Tieren, die, da sind in der Wald; Konnte man nicht alles schlachten und schickt zu uns hier in Hamburg? Wir brauchen gerade Fleisch zum essen. Es ist sehr gut dahs Ihr Sohn zurück gekommen sind; und wir hoffen ihm bessere Gesundheit.
Mein Sache, warte ich immer noch für gute Hoffnung zu bekommen von der Gericht. Es geht mir jetz nicht gut, sehr wenig Essen bekommen wir hier,

sonst werde sehr gut geht.
Die Behandlung ist sehr gut von der
Herren Aufsehren. Hoffentlich wird
bald diesem schlechten Zeit bald am ende
kommen. Man kann nicht weiter
so leben. Sonst nicht neuers.
Sende ich Ihnen u Ihre Frau viele
Grüssen mit Gehorsamst.
               Ihr Gehorsamster Junge Stephan

N. B.

Ich hoffe auch noch bald die Deutschen
die Kolonie wieder bekommen, dann
mache ich sofort nach Hause,
ich habe sehensucht von Meine
Mutter u Brüder.
               Ihr Bischoff

3.6 Stephan Bischoff, letter to Carl Meinhof, August 23, 1918. Personalakte Stephan Bischoff, 361-6-V-158, Staatsarchiv Hamburg.

Bischoff with the license to criticize, which makes for a speaking position that significantly differs from the position he inhabited when he wrote his letter to Meinhof.[56] As with Abdoulaye Niang's recording, and with the recording of many others before and after him, the moment of speaking into a gramophone became a platform (Hoffmann 2009b, 2020a, 2020b; Lange 2020; Vail and White 1991, 43).

As mentioned above, the fable (Berlin Lautarchiv, PK 862/2) was recorded in Ruhleben during a session in which terms in Ewe that were relevant to the linguistic mission had been recorded (PK 862/1). Johannes Ossey's transcription and translation reads as follows:

Gbe ɖeká gbe lá, mí kpakplé yevú áɖé míyi du áɖé
me, si wóyɔ́ ná bé krátsí.
Éye míyi duame lá, míkpɔ́ keséwó sugbɔ fũũ le duame
abé akpé atɔ̃ ené.
Yé míbíá duametɔ́wó se béna: "Nuka wɔm̀ kesé siawó
le le du sia me hãã?
Yé duametɔ́wó gblɔ béná: "Kesé lá, miáfé máwú
wònyé."
Yé yevú lá bé: "Amegbetɔ́ ményó bé né wòatsɔ́ kesé
áwɔ máwú o.
Éyatǎ yevúá vá yi tsɔ́ kesé búbu tsó aféme ési wòle
nyinyim̀, sa ka ɖé ali nê vátsí ɖé atí ŋútí.
Éye keséáwó kpɔ̀è lá, wóvá le foe.
Éyi (ési) yevúa kpɔ́ bé, wóle fo keséá éye wóle dzá
wui lá, étsɔ́ tú éye wòda kesé akpé atɔ̃, gaké lá, kpéá
méfo ɖeké o.
Éye keséawó kátã̌ wósí, éye wódzó yi gbĕme.
Ési keséáwó sí yi gbĕme lá, yevúa dzó vá yi be ɖé
xɔme béná keséáwó mágakpɔ́ yè o.
Ési keséáwó kpɔ́ bé gafofo afã̌ megbé lá, wó kátã̌
wógatrɔ́ vá.
éye duametɔ́wó ɖa nú yé wótsɔ́ ná keséláwó, yé
keséláwó ɖu. Éye wóɖu lá, wó kátã̌ wóle feféḿ. Ésia
wɔ nukúnú ná mí ŋútɔ́, béná miákpɔ́
kesé akpé atɔ̃ ánɔ anyí le amewó dome, ánɔ feféḿ
kplé ɖevíwó. Éye wóáfé ʋuu kékééké háfí wóádzó áyi
aféme.

Váséḍé égbe égbe lá, kesé siawó le du má me éye
wóyɔ́náwó bé yèwó ƒé máwú. Éyată éwɔ nukúnú gã
ŋútɔ́, béná yevúwó nákpéḍé du sia metɔ́wɔ́ ŋútí éye
wóáfíá nú wó béná wóányá nú átsó máwú ḍeká si
ameyibɔ kplé yevúwó nyá lá ŋútí.

One day I went with a white man to a place called Krachi.
At this place we came across an exceptional number of monkeys, probably around 5,000.
We asked the people: what do all the monkeys do in this area?
The people answered: the monkey is a deity to us.
Then the white man said: it is not right for humans to worship a monkey as a deity.
Then the white man went home to fetch another monkey that he had held at his home.
He leashed the monkey around his waist and to a tree.
When the other monkeys saw the stranger, they started to beat him.
When the white man saw that his monkey was in danger of being beaten to death, he took his shotgun and shot at the monkeys, but his bullets could not hit them.
All the monkeys ran away and escaped into the forest.
When all the monkeys were gone, the white man also ran away and hid in his house, so the monkeys would not find him.
After half an hour the monkeys returned and were served food by the people and ate.
After they ate they all played together.
It was miraculous to behold, how 5,000 monkeys were among the people and played with their children. They enjoyed themselves for quite a while before everyone parted.
Until today the monkeys stay at this place and are treated as deities.
This is very mysterious and should inspire the white strangers to support these people and communicate with them to learn more about the gods of black people and white people.

Apart from the practice of evangelization, Bischoff's fable comments on colonial entitlement and identifies an unwillingness to listen on the part of German missionaries, which might be as much a comment on his speaking position in the camp as it was on practices of the missionaries in Togo. In comparison with recent texts on Yeve, but also in relation to the events in Krachi (Apoh 2020; Hüsgen 2020; Ntewusu 2015; Venkatachalam 2012), the fable Bischoff presented does not interpret proselytization as a watershed moment in the process of civilization, which is how German missionaries have preferred to understand their project. Instead, the evangelizing mission and its inherent violence become an episode in the *longue durée* of the socio-religious history of the region. In the fable, a white man visits a village on the Gold Coast. Upon arrival, he arrogantly assesses the religious practices of people he has never met, telling the inhabitants that monkeys are not to be worshipped. Shortly thereafter, in clear betrayal of his initial statement, he brings his own monkey to replace the local ones. Interestingly, in the fable it is not the worshippers who fight the forcibly introduced new god; instead, it is the local deities themselves who defend their rightful place. The stranger does not hesitate in attempting to kill the deities in the place that is identified as the town Krachi. Yet they are either invincible or able to escape the bullets. Perhaps shocked by their invincibility or by the realization of his own lack of power, the white visitor retreats in fear, and the situation is restored. The story ends with the narrator's request to communicate as equals.

In the same recording session in Ruhleben, Josef Ntwanumbi inserted the Bible into a series of words in isiXhosa by means of a small alteration, turning "a book" into "The Book"—thereby adding a layer of meaning that hints at a contrast in knowledge transmission. This shift of meaning from "a book" (*ncwadi*) to "The Book" (*yincwadi*) versus "a dance" (*umngqongqo*), as a performative form of transmitting knowledge, may represent a fragment of a discourse on different ways of knowing and sharing knowledge in southern Africa, which also surfaces in recordings made by the Austrian missionary Franz Mayr in 1908 (Hoffmann 2020b; Hoffmann and Mnyaka 2015). In comparison to Josef Ntwanumbi's cautious intervention, Stephan Bischoff's critique, which was recorded after Ntwanumbi's, sounds bold, even daring. The biographical information on Bischoff, delivered by the university's files, suggests that he criticized the practices of proselytization from a position of intimate knowledge and personal experience at a mission station in the region that is now Togo and Ghana. He had grown up on a mission station—his father and perhaps also his mother were religious teachers. Because of this and because he spoke German, Meinhof's interests must have been very clear to him. Yet what made his critique particularly provocative—and timely—is the insertion of the name of a place—Krachi—where German colonial officers, twice within a decade of German colonial rule, had violently attacked the shrine of the deity called Dente and executed two *Bosomfo* (priests). The second execution took place a year or two before Bischoff left the Gold Coast. These two attempts by the German colonial power to end the worship of Dente, and also to curb the power of the shrine and the *Bosomfo* in the region, ended with their blowing up the cave in which the Dente shrine was located. The outrage this massive violation of the shrine must have caused, not only among followers, suggests that the event was discussed even in Lomé, before Bischoff had migrated to Germany.

The first execution of a *Bosomfo* of the Dente shrine, Nana Jantrubi, made it into German newspapers. In line with the presentation of the colonization of German Togoland as part of a civilizing mission, this was celebrated as an act of "freeing the natives from tyranny," writes Wazi Apoh (2020, 461). The events also entered colonial literature with Werner von Rentzell's (1922) semifictional narrative "Des großen Gottes Dente letzte Tage" (The last days of the great god Dente).[57] Von Rentzell presents the events as a story that is, in form and content, rather different from the fable as told by Bischoff.

In von Rentzell's story, "Africa" is a country; it is dark and hot and emits mysterious noises. Von Rentzell's narrative stages an imagined acoustemology of unnerving, permanently sounding difference, in which white superiority is distinguished from all that is African by its eloquent quietness.[58] He describes the young German commander of the station as a fever-ridden hero, constantly on the verge of tropical madness, whose weary soul ("ermattete Seele") listens fearfully to the "boiling darkness" of the African night. Every sound and image of this Africa is represented as threatening. The German military station is depicted as clean, quiet, and white. It is situated above a stream in contrast to the village, Krachi, which is enclosed in the hot dampness of the forest. The drums of the Dente sound indecipherable and dangerous: they are "the hiss of a beast bristling with anger" ("das Fauchen eines wutschnaubenden Tieres"; von Rentzell 1922, 61). The prayers of the priests sound like howls; the congregation shies away from the light (*tagscheu*); their polyphonic songs are wild and ecstatic. Von Rentzell's text is a string of clichés; the narrative reads like the urtext for Binyavanga Wainaina's (2019) satirical essay "How to Write about Africa."

Von Rentzell's presentation of the conflict between German colonial power and the Dente priest as entirely cultural-religious erases colonial economic interests and the concentration of socioeconomic power around the Dente shrine in Krachi. In von Rentzell's story, a fictional Christian prisoner denounces the *Bosomfo* as a murderer who sacrifices the adherents of Dente by poisoning them. The prisoner appears to be an outcast due to his Christianization; he urgently demands the execution of the Dente priest, embodying the figure of the "enlightened native" who asks the German colonizers to rescue his fellow countrymen from local tyranny. In the text, spirit language appears as secret language (von Rentzell 1922, 78) spoken by the priest with *verstellter Stimme* (feigned voice), the fearsome sound of which von Rentzell reveals to be artificially fortified by the acoustic effects of echoes in the cave. The author had traveled with the linguist Diedrich Westermann on a steamboat of the Woermann shipping line in July 1914, shortly after the second execution of a *Bosomfo*, Nana Abrakpa. Did he learn from the linguist about secret languages on the long journey to the West African coast?

Against the threatening noise of the heathen cult, von Rentzell places the quiet methods of the German military: the preconditions for successfully overthrowing the *Bosomfo*, he writes, were the swiftness and absolute quietness (*Lautlosigkeit*) of the operation (82). In his description of the strategy, one reads of whispers and low whistles. Startled by the strategy of the strangers, the African night falls silent. In the stillness of this night, the soldiers operate in silence; their stealthy action is depicted in sharp contrast to the cacophonic noise of the cult. After the silent attack, it is predictably drum language that communicates the message of the capture of the *Bosomfo* to the villages in the vicinity.

Von Rentzell's narrative interweaves delirious phantasms with historical facts; he describes the ruthless looting of all religious insignia related to the worship of Dente, which in his texts are an artfully decorated throne, a sword, and a baldachin. Yet his was not the only text that justified the looting with the aim of hindering any resurrection or continuation of the cult that troubled German plans to gain control of Krachi.[59] After the denigration of the practice of Dente worship, the narrative dehumanizes the *Bosomfo*: the colonial trial presents him as a beast, no longer human, marked by sleeping sickness as much as by alcoholism (96). Even his execution is woven into the sonic dichotomy the narrative creates: the *Bosomfo* faces a silent death by hanging. With his last words he threatens to end German occupation, and only in this scene do Hausa[60] traders, with whom the German military had actually cooperated, enter the stage of the narrative as an audience. In the story, the local chief submits to German rule (98) after the execution of the *Bosomfo*. The detonation of the shrine, which actually took place after the second raid of the cave in 1912, forms the climax of the narrative. When the shrine explodes, the forest itself seems to awaken like a gigantic animal. In a show of the power of technology, the detonation uproots giant trees and the soul of Dente flees the place, shrieking. The white man, writes von Rentzell, had now won the battle against Dente, and the black population has been saved from the tyranny of the cult by the civilizing power of German colonization (112). Von Rentzell's narrative is part of an ensemble of circulating objects and productions—texts, images, and ethnographic objects—that are directly connected to the imperial attempt to dismantle Dente worship and destroy the concentration of power around the shrine in Krachi. The archive of the MARKK in Hamburg, as well as the photographic archive of the Basler Mission, each hold a photograph of the cave after its destruction, with an unknown white man in a victorious pose at the entrance (see plate 15). The photograph was taken by Hauptman (Captain) Alexander von Hirschfeld, who had been a *Bezirksleiter* (district officer) in Togoland from 1909 to 1918.

The file to which the photograph is attached in Hamburg describes the image as the "Fetischhöhle des Gottes Dente" (The fetish cave of the god Dente). It states that human sacrifices were offered there until 1911, the last victim being a boy who was killed in that year. It also notes that both the priest and the local chief were executed by hanging. The file also alleges that the *Götzenbild* (idol) of Dente was never found. However, other objects probably made it into the collection of the Ethnologisches Museum in Berlin and might still be part of the collection of the Stiftung Preußischer Kulturbesitz (Foundation for Prussian Cultural Heritage), also in Berlin. A letter to Felix von Luschan from the colonial officer Hans Gruner, who had ordered the execution of Nana Jantrubi and the raid of the cave, speaks of plundering:

> Als ich den Fetischpriester Mossomfo in Kratschie gefangen nahm, erbeutete ich eine Reihe wertvoller Ethnografica z.B. den Fetischrock des Priesters . . . Mütze, Fetischsticks, Trommelstöcke etc. und andere Dinge.[61]
>
> Upon arresting the fetish priest Mossomfo in Kratschie I was able to capture some valuable ethnographic objects, e.g., the fetish dress of the priest . . . hat, fetish sticks, drumsticks, etc., and other things.

In line with the project of amassing objects for the Ethnologisches Museum in Berlin, which had been founded in 1873, Felix von Luschan had established networks of colonial military staff, who were commissioned to send objects to Berlin. Gruner thus informed von Luschan immediately about the desecration of the shrine.

Recent publications have reinterpreted the complex history of the shrine and its destruction by German colonial powers (Apoh 2020; Hüsgen 2020; Ntewusu 2015; Venkatachalam 2015). Meera Venkatachalam introduces the Dente shrine as an oracle where protective and clairvoyant powers were offered to political leaders of the region, to which Apoh adds Dente's assistance in farming matters, advice regarding conflicts of chieftaincy, the exorcising of witchcraft, and the function of the shrine as a safe place for refugees (Venkatachalam 2012, 130; Apoh 2020, 450–51). Samuel Aniegye Ntewusu sees the cause for the conflict between invading Germans and the *Bosomfo* as a struggle for the control of transregional trade. This conflict evolved around trade relations, specifically ferry taxes, which were a major source of income in Krachi. The *Bosomfo* taxed the riverine trade of salt, kola, palm oil, cotton, and slaves (Ntewusu 2015, 9) and thus severely curtailed German plans to extend colonial power and trade upcountry. The first execution, of the *Bosomfo* Nana Jantrubi, who was charged with human sacrifice and the harassment of the Hausa traders, was set up by Hans Gruner, who was the commander of the German colonial post Misahöhe in 1895 (Apoh 2020, 460). The trial and execution by firing squad were part of a German expedition into the Togolese Hinterland, which at that time was not controlled by the German colonial power. Jan Hüsgen (2020) describes this expedition as an attempt to extend control to the areas north of the coast. The expedition was initiated by Hans Gruner and supported by the governor of German Togoland, Jesco von Puttkamer. Hüsgen argues that the disempowerment of the *Bosomfo* in Krachi had been planned in advance as a demonstration of German power. Yet neither the execution of Nana Jantrubi nor the public hanging in 1912 of his successor, Nana Abrakpa, and his assistant, Okeowane, ended the worship of Dente (Apoh 2020, 464).

Like the words that Meinhof identified as relevant for the work of the mission, Bischoff's fable appears as an undated transcription in the files of the Lautarchiv. It is not clear, as yet, who produced this transcription. However, unlike the trophies from the cave in Krachi, the photograph that stages a victory over the Dente, and the colonial revisionist tale of von Rentzell's, this transcribed text was not met with interest, nor did it circulate among the German public. To my knowledge, it was not even discussed among experts. The fact that the archive of colonial knowledge production has swallowed this spoken text is not a consequence of its orality, versus, for instance, that of the written narrative by von Rentzell. Nor did its genre, as a moral fable that does not recount events in a form that could have counted as evidence, cause its exclusion. A debate centering on the morality of Gruner's conduct in Krachi apparently did draw the attention of German newspapers. The linguists' disinterest in or dismissal of Bischoff's fable was thus related not to its genre but to his enunciatory position (Prakash 1994, 1487). Clearly, no room was granted for the critique of a young Togolese who had experienced the practices of proselytization himself. Nor was a fragment of local orature that assumed the position of moral judgment with regard to the events in Krachi, and which questioned the evangelizing mission more generally, considered to be of any interest in Germany. The perceived subalternity of the speaker as a person of color, whose character was discussed and whose morality Meinhof was asked to guarantee, but also that of his spoken text, which was dismissed, certainly was a discursive effect in this case.

At the time of this writing, the debate around coloniality, the decolonization of institutions, and the repatriation of ethnographic objects and human remains is ubiquitous in Germany. Bischoff's fable provides no exact evidence of what happened in Krachi in 1912. Yet it presents a historical, local position that discusses those events critically (Vail and White 1991, 73). From his position as an informed contemporary critic, the fable of the 5,000 monkeys presents a conflict between different modes of worship in an abstract form, ending with the advice to communicate as equals.[62] In this concise text, the event becomes an episode in a long history that has seen the appearance and disappearance of competing religious cults around which social alliances have been forged and negotiated. The text Bischoff chose to speak into the gramophone funnel presents proselytization as a moment of crisis in which a social and religious order is disturbed yet neither destroyed nor permanently transformed. Instead of a grand narrative of development toward Christianity, Bischoff's fable gestures toward complex, entangled histories of spirituality and worship in which the evangelizing mission is but one episode.

Stephan Bischoff probably left Germany to return to Togo in 1919.

Fragment V

Mamadou Gregoire: "The sea requests fish from the rivers"
TRANSLATED FROM GERMAN BY ANETTE HOFFMANN

The sea requests fish from the rivers
Which is not fair, because the sea is bigger.
In the same manner, the king requests tribute,
which he should not because he is more powerful.
In this way, King Buasik [Béhanzin?] was too
    demanding.

(Phonogramm-Archiv Berlin, Phon. Komm. no. 120)

Mamadou Gregoire was a *tirailleur sénégalais* who came from Ouidah, a coastal town in what now is Benin. When Hermann Struck drew his portrait in the Wüns-

3.7 Hermann Struck, *Portrait of Mamadou Gregoire*, undated. From Struck 1917, plate 98.

dorf camp, Mamadou Gregoire must have been thirty-two years of age (see figure 3.7). His text, poem, or song, which I accessed as a handwritten transcription in German at the Phonogramm-Archiv, refers to the recruitment of men for the army, most probably during the reign of King Béhanzin of Dahomey (1888–94). Gregoire's text, probably translated from French to German by the musicologist Georg Schünemann, might be drawn from an oral repertoire of critique that commented on the recruitment of soldiers during the colonial wars between France and the kingdom of Dahomey (1890 and 1892–94), during which *tirailleurs sénégalaises* fought for the French army against the army of the kingdom of Dahomey. Perhaps Mamadou Gregoire's poem gestures toward a relativization of enmity during contemporary war, World War I, in which he fought for a colonial army and became a prisoner of war in Germany.

Afterword

KNOWING BY EAR

Most of the records in the Berlin Lautarchiv remained cloistered silently in their depositories for almost a century. As one of their first critical listeners, Britta Lange perturbed the archive's long hibernation and also sparked my research into the recordings with African POWs. My engagement with these recordings, which result from the most systematic project of linguistic recording conducted in Germany in the twentieth century, pushed me to investigate more systematically how voice recordings may speak of and to colonial pasts, how subaltern enunciative positions may surface acoustically, and how the acoustic archival trace of speakers may differ from both their visual traces and from how they were represented in writing.[1]

Listening closely as well as collectively—together with translators, historians, and linguists—I have attended to what voice does and how genres and particular lingual practices operate. The history of the making of the Lautarchiv cannot be told without these recordings. Its telling demands that critical attention be paid to the generative power of colonial knowledge production, the epistemic violence ingrained in the project, and the ways in which the spoken and sung texts of captured foreign detainees were objectified as archival artifacts. Listening to these acoustic sources of history has meant attending to the fabric of the Lautarchiv and to the processes of knowledge production with POWs in Germany. My attempt to know by ear has impelled me *not* to understand these recordings as "interesting raw material" that can be mined from the archive in the form of sounding objects, which still seems to be the most common contemporary use of archival sound. Instead, I have listened to and engaged with speakers' utterances as profound yet often fragmented aspects of historiologies and as echoes of discursive formations, for instance, in the cases of the songs that Abdoulaye Niang carried with him from Dakar and of the snippets of a critical conversation about the

Dervish Movement in Somalia that traveled to Berlin with Mohamed Nur. Our hearing and reading of these voiced elements of biographies, aspects of journeys, and the fragments of histories that the speakers chose to address made for the kaleidoscopic texture of the chapters in this book, which are arranged around speakers and their recordings, and not around recordists or a chronology of the Lautarchiv.

Because these performed texts and utterances are not raw material, I have sought to engage with their spoken or sung content, paying attention to their speakers' enunciative positions vis-à-vis the epistemic practices of colonial linguistics that prompted these utterances. I have done so by listening with experts in the languages spoken, and by engaging with the recordings' performativity, their different genres, their uses of metaphor, and the ability of voice to communicate beyond words. My attempt to "know by ear" has taken seriously how the productive power of colonial knowledge production has watermarked the recordings by labeling them without a sense of or interest in their semantic content, and by organizing them according to ethnic groups and languages. This has also meant holding in tension, throughout this book, the epistemic practices and racializing concepts of German linguists that informed the production and the archival order of things in the Lautarchiv, and the variety of genres, prosodic practices, and ways of speaking of and from the (now) historical subaltern positions that are found in this archive.

The recordings of the Lautarchiv have allowed me to listen in on knowledge production in German POW camps, a process that aimed to generate artifacts in the form of recorded example sentences for languages that were to be studied at a later date but that mostly never were. Listening in on the making of these linguistic recordings has allowed me to hear the recorded voices of African soldiers and migrants speaking of World War I, with a sense of belated "earwitnessing." But in contrast to what the notion of earwitnessing might suggest, one does not hear World War I in these files. Instead, what one does hear are its repercussions in spoken and sung texts. These echoes resound in many of the recordings of prisoners who spoke or sang of this war or reminisced about the circumstances that had led to their detainment in camps. To my mind, Aimé Césaire's (1990, xliii) notion of the contrast between "half-starved scientific knowledge" and the transformative potential of poetry resonates strongly with the recordings and their documentation, which turned even the most dramatically voiced appeals and poetic interventions by prisoners into decontextualized language examples. Epistemic and archival practices swallowed words and made spoken texts disappear in the written record.

Listening in on colonial knowledge production and hearing these recordings as sources of history was a process that initiated what Carolyn Hamilton (2011) has called the "re-deeming" of archive(s) at a particular moment in time or over a period. It has altered what we expect the Lautarchiv to contain, as much as how we approach it and how we position ourselves as listeners (Lange 2020, 2022; Hilden 2022, 2015; Macchiarelli and Tamburini 2018; Riva 2015). Attributing a different status to the recordings has also considerably altered the position of the archive itself—from a depository for which I had to pick up the keys to be able to let myself in, in 2011, the Lautarchiv has advanced to become a quite prominent part of the collections at the Humboldt Forum, even featuring in the permanent exhibitions at the Forum.

What Resounds or Does Not Resound in Public(s)

Some historical eras seem to lend themselves to being known by ear, while others remain almost eerily silent. Whereas the fascist era in Germany left an echo, in the sense that it is recalled sonically as well as visually, as Carolyn Birdsall (2012) has shown, the archival trace of the colonial era in Germany is mostly visual or appears in writing. This book attempts to show that colonial knowledge production did, in fact, leave a partial soundtrack in places that include the Berlin Lautarchiv and Phonogramm-Archiv. Like Birdsall (2012) in her critical acoustemology of Nazi soundscapes, I advocate critical listening to enable us to develop a multifaceted historical understanding, and to attend to, as well as engage in, "earwitnessing."[2] Birdsall's attention to acoustic memories and her discussion of earwitnessing put the notion and also the history of witnessing as visual into perspective. The historical echo her book conjures is productive beyond its direct engagement with German fascism, because it contributes to a historical practice of listening combined with close attention to discursive formations and visual practices. While the collections of the Lautarchiv add up to a partial soundtrack of colonial knowledge production in Germany, these recordings, in contrast to, for instance, some of the photographs that were taken in the camps, were not created for public consumption. Without translation or contextualization of their content and genres, the recordings that have been played in public places such as museums have mostly featured as ambient sound.

An example of the effects of the belated release of hitherto unheard speaking positions into the public is a conversation that took place in January 2019 in the space of my exhibition *Der Krieg und die Grammatik: Ton- und*

Bildspuren aus dem Kolonialarchiv (War and grammar: Audiovisual traces from the colonial archive) at the MARKK in Hamburg (see plate 16).[3] During a walkabout, one of the visitors commented angrily on the sound installation in the first room, in which Albert Kudjabo's drumming of "war" preceded a well-known historical recording of the German Kaiser's declaration of war, and in which Wilhelm II's speech was interrupted by Jámafáda's comment on that same war, in which he fought as a soldier in the French army. Jámafáda's response to the Kaiser's declaration of war marks the asymmetric nature of their speaking positions as well as the unequal measure of attention they have attracted in the course of the intervening century. The delayed release of the recordings of Albert Kudjabo, Jámafáda, and Abdoulaye Niang into the public space of a museum directs the visitors' attention to the question of who is positioned to speak (of) history. It does so not only by making audible that one speaker declared a war in which the other had to fight, but also by reminding the listener that the Kaiser's speech is familiar (for instance, from documentary movies) and citable, whereas the prisoners' voices are appearing publicly in Germany for the very first time. The visitor who approached me during the walkabout made clear that he did not appreciate what he conceived of as the messiness of an installation of acoustic traces from the Lautarchiv. Nor did he like the way I had staged a dispute between recorded voices that did not, and could not, have taken place in 1917. Yet this, as I suggest in this book and with my ongoing curatorial work, is what listening to the Lautarchiv (and to other colonial sound archives) can facilitate. In felicitous cases, listening to the colonial archive interrupts and questions the persistent reverberation of entrenched Eurocentric narratives and imagines an arrangement of what could have been done differently.[4] For Samatar Hirsi Egeh, who also visited my exhibition, hearing historical chants in Somali in a German museum triggered personal memories. He had come to Hamburg to tell Bodhari Warsame and me about his family's history, which was connected to Mohamed Nur, who was the focus of the exhibition. In the café of the museum, he recalled orally transmitted history: that Nur had been the teacher of Hirsi Egeh Gorseh's children in the ethnic shows that Egeh Gorseh had organized together with Carl Hagenbeck in Hamburg.

With another visitor at the same exhibition, this time at the opening, I was afforded the chance to discuss the question of what can be gained by intuitive, perhaps uninformed, listening alone. The professor emeritus of the University of Hamburg took issue with my curatorial choice to stage Fatou Cissé Kane and Johannes Ossey's spoken presentation of selected recordings that they translated into German. Arguing in favor of this curatorial strategy,

I mentioned that my idea had been to take the recordings of Kaiser Wilhelm II (who declares war), Wilhelm Doegen (who speaks of recording opportunities presented by the POW camps), and the philologist Heinrich Lüders (who speaks of recording opportunities presented by the POW camps and stresses the importance of orientalist research in German) and juxtapose them with the original recordings of Jámafáda (in Mòoré), Abdoulaye Niang (in Wolof), and Josef Ntwanumbi (in isiXhosa). In my first conception, the POWs would have spoken in their original languages and the translations and transcriptions would have appeared only as texts on the walls. Yet in conversations with Jos Thorne, the exhibition designer of *War and Grammar*, it became clear that had I done so, the installation would have orally and aurally repeated the durable trope of unintelligibility, or even the perceived obscurity of African languages and texts. Kaiser Wilhelm II, Doegen, and Lüders would have been understood by German-speaking visitors, while the untranslated historical recordings of Jámafáda, Abdoulaye Niang, and Josef Ntwanumbi would have been positioned, again, as ambient sound. Consequently, I opted for a mix of original recordings—in German, Mòoré, Wolof, and isiXhosa (in the first room)—together with translated texts, spoken by Fatou Cissé Kane and Johannes Ossey. Translation, I added, can be a strategy for extending what is available in the present as arguments and texts, which is an attempt to undo the aura of the exotic, strange, and unknowable (Spivak 2010). Apparently, this was not convincing. The visitor told me that he would have preferred to immerse himself in the sound of the original recordings, and that he believed he would have been able to make sense of the mood of the spoken languages without our translations. To my question of whether he would expect to discern the weather report in Japanese from the rest of the news in Japanese, without visuals and while not understanding the language, he did not respond.

Seeking to know by ear has also sharpened my awareness of my own position as a listener—my privilege of access to this archive in Berlin—and the continuing absence of these recordings in the countries whence the speakers came. In attending to listening positionalities collaboratively, I benefited immensely from the work of Jennifer Stoever (2016), Maria Ochoa Gautier (2014), Anne Storch (2011), and Dylan Robinson (2020). Their work, together with the insights that came with translation, sharpened my awareness of the systemic racism (as related to voice) in the publications of the musicologist Fritz Bose (1943/44), in the comments that Wilhelm Doegen added to the personal files of the men he made recordings with, and in prevailing colonial attitudes that gave German listeners the sense that they were entitled to judge performances they did not understand. Close listening meant "staying with the trouble" of epistemic violence while at the same time not burying the potential of the speakers' often opaque forms of critique and performative expression in its shadow (Haraway 2016). This said, staying with the trouble while continuing to listen also entailed not reducing historical sound recordings entirely to the disciplinary imperatives that organized their archiving. It also taught me, again, to factor in, but not to overestimate, the power of colonial linguistics, which suspended communication in the moment of recording. Close listening makes clear that this suspension of communication was one-sided: POWs continued to speak, and at times their audiences were other prisoners, and not the linguists. Nor should their songs and performed text be heard as having been created by the archive, even if it was the desire to archive them that had prompted their presentation. The musical pieces and the performed texts do not belong to the Lautarchiv, even if it holds them in conserved form and despite the fact that European copyright has orchestrated this understanding. Instead, they belong to a much larger cultural repertoire and heritage, which may still be able to embrace their meaning.

Instead of merely "mining the archive" for interesting material, informed reading and listening can make an audible difference to musical and artistic practice (Hamilton 2013). This became clear in Heiner Goebbels's composition *House of Call*, which premiered at the Berlin Philharmonic in 2021. The composer had contacted me in 2010. His idea was to create music with the historical sound recordings Hans Lichtenecker had produced in Namibia in 1931, on which I have written extensively and on the basis of which I curated the exhibition *What We See* (2009–10).[5] Heiner Goebbels particularly liked the recording with a young Khoekhoegowab speaker, who appeared in Lichtenecker's documentation under the name Haneb. While the recording may sound like a joyful song, the translation, made in 2008, makes clear that Haneb actually sings of danger and crisis (Hoffmann 2014b). The song Haneb presented for the recording on a German farm where he worked, in what was then South West Africa, reenacted the situation of the Khoekhoegowab-speaking protagonist of a story: the boy, whose name is !Hansoroxatsoab, tries to warn his small community against imminent danger. In a more habitual setting of storytelling — one that is not framed and constrained by the epistemic violence of anthropometric measurements and the making of life casts in which it actually was presented and recorded — this song would have formed the center of a story. In the story, !Hansoroxatsoab, whose social position is described as marginal or precarious ("sitting with his butt in onion peel"), involuntarily witnesses a planned attack, while everyone else is not at home (Hoffmann 2014b; Schmidt 2001). In the story, the singer's social position does not allow him to address the others directly. Unable to warn them immediately, !Hansoroxatsoab's strategic response is to communicate what he has witnessed in the form of a song, which enables him to present his knowledge in a way that entices others to listen to him. When they lend an ear to his pleasant tune, they realize that he is trying to warn them. My understanding of this song and the position of Haneb, who embodied !Hansoroxatsoab's position, is based on the translations of Namibian historian Memory Biwa and Namibian linguist Levi Namaseb, as well as on a published version of the fable that usually frames the song. I also drew on my research into Lichtenecker's project of anthropometric documentation, and on what can be known of the history of black farm workers on white farms in Namibia at that time. In my reading, Haneb impersonated the boy who is not in a position to speak. As a black farm worker, Haneb was subjected to Lichtenecker's project of creating racializing images of people of color. Haneb's song stresses the importance of listening and cites a known theorization on unequal speaking positions, which the story transmits. Archived as an example of music in the Berlin Phonogramm-Archiv, the recorded song had not been translated before 2008. My research that presented the results of close listening and of reassembling a collection that had been dispersed across several institutions in both Namibia and Germany, together with the availability of this published work, which resulted from cooperation between German and Namibian researchers, arguably allowed the composer to approach the song with an informed ear. Based on his long-term engagement with this song and the collection, Goebbels created an orchestral piece that attentively weaves sound around Haneb's song. The composition, played by the Ensemble Modern, conducted by Vimbayi Kaziboni, seems to reach out to Haneb, if only retrospectively. The orchestra does not drown the song out, nor does it exoticize it. To my ear, Goebbels's musical engagement did not suffocate but rather embraced this particular recording and its double meaning as well as the historical circumstances of the recording moment.

In this book I have followed the archival traces of the speakers as far as possible, and I have focused on their positions as speakers, narrators, migrants, and prisoners and also as historians, authors, poets, and makers of the Berlin Lautarchiv. Stressing their contribution to this archive, I argue that this archive is not the result of what was then the exciting new technology of phonography harnessed by ambitious and innovative researchers, which was how the Lautarchiv's endeavor was presented at the Humboldt Box (see the prologue). Instead, I suggest that the project of recording with hundreds of prisoners in German internment camps should be understood as the result of war, imprisonment, and the opportunism of researchers, some of whom had previously practiced extractive knowledge production in colonized countries. Moreover, even if the recordings were created under these constraints, the authorship of hundreds of incarcerated men who told stories, sang songs, narrated, counted, made music, presented drum language, and recited texts still needs to be acknowledged.

While the work of re-personalizing the speakers has informed the writing of this book, close attention to content, form, genre, and authorship of historical recordings is also a decolonial strategy to undo, again incompletely, the *chosification* (Césaire 2017) of spoken or sung texts and pieces of music that have become archival objects. Listening closely to what was communicated or told ruptures colonial knowledge production's denial of the central contribution of the performers. This denial has historically benefited archivists and the archive by attributing to them the sole authorship and copyright for these documents. Yet neither this nor any other strategy will decolonize the Lautarchiv.

Another decisive aspect of the colonial history of the Lautarchiv (and other sound archives in Europe) is their location. After decades of critical engagement with colonial archives in the academy and more than one hundred years from the time of their production, these recordings are still mostly inaccessible in countries where the speakers and singers came from. In other words: systematic restitution still has to happen. This means that despite our critical ear(s) and those of others who have listened to the recordings in Berlin, we have done so from a privileged position. Several of the researchers who have engaged with this archive have articulated their discontent with the continuing inaccessibility of the recordings outside of Germany (Garcia 2017; Hoffmann 2015a, 2020; Kalibani 2021). Furthermore, Britta Lange (2020) has critically marked her positionality as a privileged listener on the CD that accompanies her monograph *Gefangene Stimmen*.

Whereas Heiner Goebbels, as a well-known German composer, was enthusiastically invited to engage with the recordings by the Berlin Phonogramm-Archiv, such easy access was not granted to meLê Yamomo, a musicologist and performance artist based in Berlin and Amsterdam. Several of his requests to access the recordings of the Phonogramm-Archiv were left unanswered. Yamomo is unlikely to be the only composer and musicologist who experienced gatekeeping by a sound archive in Europe. His work *Echoing Europe* (2019), which has recently been shown at the Ballhaus Naunynstrasse in Berlin, criticizes the current politics of access together with Western musicologists' historical approaches to non-European music. His performance stages the persistent chauvinistic practices of judging what does or does not count as art in Europe, which has an ongoing influence on who is or is not granted access to sound archives in Germany. Apart from his work as a sound artist and theorist, Yamomo's project, "Decolonizing Southeast Asian Sound Archives,"[6] supports and funds the research and artistic engagement of non-European sound artists, musicians, composers, and theorists whose musical heritage is locked in European archives. In this way, a temporary project that is funded by the European Research Council takes on the responsibilities of the colonial trace that should be accepted by the institutions in Europe to which the sound archives belong.[7] By "taking on the responsibility of the trace," I mean making known the results and documentation of colonial knowledge production and systematically, proactively, organizing the restitution of these recordings.

This said, promoting restitution is not synonymous with decolonizing the archive, nor should it result in passing on to others the responsibility and curatorial care of the recordings and their documentation. Neither artistic interventions nor restitution will exorcise the history and imprint of violent epistemic practices from the colonial archive. Yet specifically with regard to sound recordings in their digitized form, it is possible to keep *and* restitute the acoustic documents, which means that these recordings can speak and be heard in several places. As a strategic move, this could create opportunities for collaborative critical listening and for understanding colonial histories from different perspectives or listening positionalities (Robinson 2020). Digitized sound recordings from the colonial archive can and need to travel between the archive, places where they were recorded, and the places where the speakers and singers once came from. I have argued in this book that listening to them in Europe as well as in the countries where the recorded languages are understood by more listeners alters our understanding of what is to be found in the colonial archive. It may also enable listeners to find versions of history that are not archived elsewhere. These versions of history need to enter public spaces to speak, to resonate, and to communicate the acoustic testimonies of witnesses to colonial history, their narratives, critical interventions, and fragments of historiology.

....

Some of the recordings from the Berlin Lautarchiv that are presented in this book have traveled already; but they need to travel farther still. In the fall of 2022, Fatou Cissé Kane and I traveled to Dakar. With the permission of the director of the Berlin Lautarchiv, Christopher Li, we discussed the modalities for a digital restitution (or sharing) of the African recordings with colleagues at the Institut fondamental d'Afrique noir (IFAN) at the Cheikh Anta Diop University in Dakar. At this point we await the decision of a funder that will hopefully support our effort to translate the documentation from German and conduct more provenance research for the "sonic return" of these recorded voices in 2024. My sound installation on the basis of the recordings that are discussed in this book has been shown under the heading *Foreign Subjects* at the Bergen Assembly (Norway, 2019).[8] *Foreign Subjects* combines the recordings of Mohamed Nur, Asmani ben Ahmad, Shire Rooble, Jámafáda, and Josef Ntwanumbi. All five men speak of their journeys to Europe and their experiences as migrants or as soldiers in World War I. More installations have been produced already; they are waiting to travel once the process of restituting all recordings with African speakers to Dakar has begun.

In recent years, historical sound archives have featured in the work of several artists, musicians, and filmmakers in Germany and in southern Africa: Nashilongweshipwe Mushaandja, meLê Yamomo, Memory Biwa and Robert Machiri, and Britta Lange and Philip Scheffner, to name but a few. In addition, archival recordings relating to Africa and Africans in World War I were featured in the opera *The Head and the Load* (2018), a collaboration of William Kentridge, Philip Miller, Thuthuka Sibisi, and Gregory Maqoma, performed in England, Germany, the Netherlands, and the United States.[9] The works of all these artists and others have already released echoes of colonial voice recordings into a wider, international public sphere. These sonic interventions introduce alternative ways of presenting, attending to, and speaking about colonial pasts as listeners and as belated earwitnesses. They allow us to hear the fragmented echoes of colonial history and to engage critically with the experience of *knowing by ear*.

Acknowledgments

The research for this book has been supported by numerous intensive collaborations across continents, which I appreciate profoundly.

I am deeply grateful to the translators of the sound recordings, who were also listeners, analysts, and interpreters, and who shared their knowledge with me generously. With Phindezwa Mnyaka, I wrote "Hearing Voices in the Archive" (2015). Serigne Matar Niang searched for traces of Abdoulaye Niang in Dakar and on the island of Goreé. Fatou Cissé Kane and Johannes Ossey were speakers in the exhibition *War and Grammar* (2019). Constanze Kutsch Lojenga took the recordings to the Democratic Republic of Congo (DRC) and arranged for the translations of Gilbert Katanabo Muhito and Faustin Sambu Avetsu, whom I unfortunately could not meet in person. Dishon Kweya was my colleague in Cape Town, where we had many conversations about the recordings. With Bodhari Warsame, I presented our work on Mohamed Nur and on ethnic shows in Germany at the Humboldt University in Berlin; we also had a public discussion on my exhibition, together with Gabriel Schimmeroth, at the MARKK (Museum am Rothenbaum — Kulturen und Künste der Welt), Hamburg, in 2020. Odette Kudjabo made time for many conversations about her late grandfather Albert Kudjabo and also shared family photographs with me. This book could not have been written without their insights and intellectual involvement.

Earlier drafts of every chapter of this book were discussed in the research workshops at the Archive and Public Culture Research Initiative at the University of Cape Town. My work benefited substantially from the discussions and productive critique of my colleagues there.

Both my research and my curatorial work received generous funding from the Mellon Foundation (USA) at the University of Cape Town; from the National Research Foundation (South Africa) at the University of Fort Hare; from the Fonds zur Förderung der wissenschaftlichen Forschung (Austrian Science Fund) at the Akademie der bildenden Künste in Vienna; from the Fritz Thyssen Foundation in Germany at the Institute for African Studies and Egyptology at the University of Cologne; from the Kulturstiftung des Bundes for my exhibition at the MARKK in Hamburg; and from funds of the Leibniz

Preis, generously disbursed by the prizewinner Anne Storch at the Institute for African Studies and Egyptology in Cologne, which also contributed to the publication of this book.

I am intellectually indebted to several academic institutions: the SARCHi Chair at the University in Fort Hare, the Archive and Public Culture Research Initiative at the University of Cape Town, the Akademie der bildenden Künste in Vienna, and the Institute of African Studies at the University of Cologne.

Finding the traces of prisoners of war in Europe would not have been possible without the support of several archives, museums, and libraries: the Lautarchiv in Berlin, the Phonogramm-Archiv in Berlin, the archive of the Department of Evolutionary Anthropology in Vienna, the Staatsarchiv in Hamburg, the Archives générales du Royaume et Archives de l'État in Brussels, the Deutsches Historisches Museum in Berlin, the archive of the Frobenius Institut in Frankfurt, the Staatsbibliothek in Berlin, the Museum Kunstpalast in Düsseldorf, the MARKK in Hamburg, the library of the African Studies Centre in Leiden, and the Historical and Special Collections of the Harvard Law School in Cambridge, Massachusetts.

Colleagues and friends in South Africa, the Netherlands, Germany, Austria, and elsewhere supported my work over many years. For their invaluable engagement and contributions I thank the following individuals: Max Annas, Abdudd-Dayyaan Badroodien, Margit Berner, Carolyn Birdsall, Micho Bücher, Michael Bull, Mbongiseni Buthelezi, Hans Christ, Marcel Cobussen, David Cohen, Marleen Dekker, Tahir Della, Ana Deumert, Iris Dressler, Jannik Franzen, Jochen Hennig, Anne Folke Henningsen, Dag Henrichsen, Irene Hilden, Samatar Abdi Hirsi, Duane Jethro, Guno Jones, Rifqah Kahn, Bodhisattva Kar, Alirio Karina, Kerstin Klenke, Christian Kopp, Annette Kraus, the members of the Kunci Study Forum and Collective, Christoper Li, Heike Liebau, Athambile Masola, Katarina Matiasek, Birgit Meyer, Renate Meyer, Thokozani Mhlambi, Gary Minkley, Nicholas Mirzoeff, Wayne Modest, Litheko Modisane, Andreas Netzer, Sandile Ngidi, Michael Nixon, Barbara Plankensteiner, Debra Pryor, Clemens Radauer, Himal Ramji, Regina Sarreiter, Johanna Schaffer, Sophie Schasiepen, Gabriel Schimmeroth, Katharina Schramm, Nick Shepherd, Katleho Shoro, Anthony Sloan, Reinilde Sotiria König, Holger Stoecker, Jennifer Stoever, Ferdiansyah Thajib, Jos Thorne, Carl Triesch, Hedley Twidle, Karl Wedemeyer, Albrecht Wiedmann, Catharina Winzer, John Wright, Mduduzi Xakaza, meLê Yamomo, Ben Zacharias, and Niklas Zimmer.

The manuscript benefited immensely from the comments of the two reviewers; from many years of discussion and intellectual exchange with Britta

Lange; from earlier readings of draft chapters by Caroline Hamilton, Esther Peeren, Ruth Sonderegger, and Anne Storch; from Jonathan Sterne's firm encouragement to write it; from Saskia Lourens's thoughtful editing of the prologue; and from Rosemary Lombard's conscientious, sensitive editorial work, along with our discussions during the entire writing process.

At Duke University Press, the entire process of reviewing and preparing the manuscript for publication was gently guided by Courtney Berger, and it was carefully edited by Livia Tenzer. I owe all of you many thanks.

Notes

PROLOGUE: CATCHERS OF THE LIVING

1 See, for instance, Mangin 1910; Fogarty 2014; Lunn 1999; Diallo and Senghor 2021; and Diop 2014, among others.
2 On the World War I correspondence between Senegalese soldiers and their friends back home, see Descamps et al. 2014. See also Bakary Diallo's book, *Force Bonté* (1926), and the writings of Lamine Senghor, which have been republished with annotations and an introduction by George Robb (Diallo and Senghor 2021).
3 There have been several recent publications on the SS *Mendi*, and efforts to engage its history. These include a conference at the Centre for African Studies at the University of Cape Town in 2017; the play *Did We Dance: Ukutshona ko Mendi*, written by Lara Foot and directed by Mandla Mbothwe, shown at the Baxter Theatre in Cape Town (2012); and Fred Khumalo's novel *Dancing the Death Drill* (2017). See also "SS *Mendi*," *South African History Online*, accessed June 6, 2023, https://www.sahistory.org.za/article/ss-mendi.
4 In this book I mostly refer to the names of the POWs as I have found them in the Lautarchiv. Many, if not most, of the names were distorted in the writing of the German linguists. I have adjusted only three names: Abdoulaye Niang, whose name was (incorrectly) written as Abdulaye Niang; Mohamed Nur, who appears as Muhammed Nur in the writings of the Maria von Tiling and in the files of the Lautarchiv; and Josef Ntwanumbi, who appears as Josef Twanumbee in the Lautarchiv.
5 The Humboldt Forum (https://www.humboldtforum.org/de/) is a museum in Berlin that now houses the collections of the former Ethnological Museum of Berlin and Museum of Asian Art, together with two sound archives, the Berlin Lautarchiv and the Berlin Phonogramm-Archiv. Opened in 2021 and located on Museum Island, the building's facade replicates that of Berlin's historic Prussian castle (Berliner Schloss), which was demolished in 1950 by East Germany, after being damaged during World War II. The new museum has been massively criticized—beginning with controversy over the political implications of mimicking a Prussian castle on the grounds of the building that replaced it, the Palast der Republik, used for cultural and political events by East Germany, and itself demolished in 2002. More recently, a critical debate addressed the colonial history of the enormous ethnographic collection and the flimsy, often contradictory, and at best uninformed concepts used in presenting these objects in the Humboldt Forum. At the time of writing, the public discussion around the Humboldt Forum—which was initiated by activists organized under the umbrella group No Humboldt 21—continues. The debate revolves around the politics of repatriation, the

rightful ownership of artworks and human remains from formerly colonized countries, curatorial practices, and the inclusion/exclusion of the societies from which these objects were taken (or stolen) in decisions related to those collections. See, for example, Hilden, Merrow, and Zavadski 2021; Förster 2010.

6 Now housed at the Humboldt Forum, the Lautarchiv of Humboldt University in Berlin holds recordings dating as early as 1909. For an official history, which does not credit the speakers discussed in this book as creators of the archive, see https://www.lautarchiv.hu-berlin.de/en/introduction/history-and-perspective/.

7 The new exhibition on the Lautarchiv at the Humboldt Forum, which opened in 2020, does not fundamentally alter the approach that foregrounds the historical invention of the recording device and the aims of the Royal Prussian Phonographic Commission for understanding the Lautarchiv's history. Britta Lange's work and my work on sound archives are quoted prominently on the walls in a section of the exhibition that speaks of the Humboldt Forum's collection. Our request to focus on the content of recordings, the speakers, and the methods of the production of the sound collection does not feature in the museum.

8 The postcard bears the caption, "Der erste kriegsgefangene Kongoneger (auf dem Transport von Namur nach Deutschland)" (The first Congolese negro prisoner of war [on the transport from Namur (Belgium) to Germany]). See *Dortmund postkolonial*, January 17, 2016, http://www.dortmund-postkolonial.de/?attachment_id=4100.

9 Paul Panda Farnana's experiences in Germany appear through the lens of Willy van Cauteren's memoir (1919). For practices of depicting African POWs, see Burkard and Lebret 2015.

10 Anke Nehrig's translations (see Fragment I) were produced for the Berlin Phonogramm-Archiv and were kindly given to me by the archive in exchange for other translations. The translations from Mòoré (see Fragment II) were created by a refugee in Germany who wants to remain anonymous.

11 On knowing by ear as acoustemology, see Feld 1996, 2015. On knowing (imperial) history by listening, see, for instance, Birdsall 2012; Ochoa Gautier 2014; Morat 2014; Missfelder 2012; Smith 2001; Rosenfeld 2011; and Hirschkind 2006. On technologies of hearing and modern cultures of listening, see Sterne 2003, 2015; Schmidt 2000; Rice 2015; Nancy 2007; Bijsterfeld 2008; Erlmann 2010; Mhlambi 2008; and Bull and Cobussen 2021, among others.

12 See "Der Krieg und die Grammatik: Ton- und Bildspuren aus dem Kolonialarchiv," MARKK, accessed June 6, 2023, https://markk-hamburg.de/ausstellungen/der-krieg-und-die-grammatik/.

INTRODUCTION: LISTENING TO ACOUSTIC FRAGMENTS

1 Many of the names of persons recorded for the Lautarchiv were misspelled. Throughout this book, I use spellings that follow current-day accepted standards that reflect the naming and spelling practices of the men's respective languages.

2   For new (or recent) work on archiving sound, see the following: Ajotikar and van Straaten 2021; Birdsall and Tkaczyk 2019; Bronfman 2016; Garcia 2017; Kalibani 2021; Robinson 2020; Yamomo and Titus 2021; Yamomo 2020; Hoffmann 2020a; Lange 2020.

3   This does not hold true for all early recordings. In his 1904 article "Einige türkische Volkslieder aus Nordsyrien und die Bedeutung phonographischer Aufnahmen für die Völkerkunde" (Some Turkish folksongs from northern Syria and the meaning of phonographic recordings for ethnology), Felix von Luschan delivered translations of the songs he had recorded in 1901. For him, the texts and meaning of the content were important. Von Luschan writes, "Leider bin ich nicht in allen einzelnen Fällen sicher, den eigentlichen Sinn jedes Liedes richtig erfasst zu haben" (Unfortunately, I am not sure in all cases whether I have grasped the actual sense of the song) (1904, 183). Ironically, many of the "Turkish songs," which were recorded during his work on an archeological project in the Middle East, were sung by an Armenian boy. Unless otherwise noted, all translations from German in this volume are my own.

4   For information on the film, see "Boulevard d'Ypres/Ieperlaan," Centre Vidéo de Bruxelles, accessed September 22, 2023, https://cvb.be/en/movie/boulevard-dypres-ieperlaan.

CHAPTER 1. ABDOULAYE NIANG: VOICE, RACE, AND THE SUSPENSION OF COMMUNICATION IN LINGUISTIC RECORDINGS

1   The exhibition at the Humboldt Box was called *[laut] Die Welt hören* ([loud] Listening to the world) (2018). From 2011 to 2019, the Humboldt Box served as a showcase for the Humboldt Forum, then under construction.

2   The Odeon Lindström Company was founded as International Talking Machine Company in Berlin in 1903. By 1906, it had more than 10,000 records, including (then already) so-called World Music, in their sales catalogues.

3   For more on the Berlin Lautarchiv and on practices of recording in German POW camps, see Lange 2011a, 2011b, 2015, 2020; Hilden 2015, 2021; Kaplan 2013; Macchiarelli and Tamburini 2018.

4   On the question of the standardization of languages, which includes the creation of entities such as languages or dialects and was related to colonial politics with regard not only to epistemological practices but also to territories of influence, see, for example, Deumert and Mabandla 2018; and Irvine 2008, among others.

5   Serigne Matar Niang is not related to Abdoulaye Niang. He lives in Cape Town and kindly agreed to translate the recordings in 2013. His engagement with the recordings of Abdoulaye Niang led him to search for relatives of Abdoulaye Niang on the island of Gorée, which lies off the coast of Dakar, Senegal, in April 2018. So far, traces of Abdoulaye Niang's family have not been found.

6   See, for example, the essays in Das 2011. See also Eschenberg 1991; Fogarty 2008; Roy, Liebau, and Ahuja 2011; Hoffmann 2014a; Diop 2019; and *1914–1918 Online*:

International Encyclopedia of the First World War, https://encyclopedia.1914-1918-online.net/home/.

7   The conceptual term *versioning* as a verb stems from Esther Peeren's work on popular culture. Peeren (2009) describes *versioning* as a decentralizing practice that allows for the alteration of texts within different performances. Thus, although versioning indicates continuity that, in the case of oral texts, is linked to the recognition of genres, as a verb it implies the processuality of altering spoken texts, which is related not only to the situative context and the position of the speaker but also to the audience, the location, and the performative frame. For historiography, versioning could mean opening the discourse for oral versions of history from other positions and locations.

8   See Diallo [1926] 1985; Lunn, 1999, 2011, 1987, 28–53; Koller, 2011b.

9   Transcribed and translated by Serigne Matar Niang.

10 An exception was the recruitment campaign led by Blaise Diagne, who successfully convinced more than 60,000 of his compatriots to enlist with the prospect (for those who came from the four communes in Senegal) of gaining French citizenship for their services. See Echenberg 1991, 365; Lunn 1999, 42–44; and Koller 2014, 2017. Blaise Diagne (1872–1934) was the first Senegalese to be elected a member of the French Parliament.

11 A detailed instruction for recording with a phonograph had been added to the manual for "ethnographic observation and collection" in 1904 (*Anleitung für ethnographische Beobachtung und Sammungen in Afrika und Ozeanien*, published by the Königliches Museum für Völkerkunde, Berlin). The first version of this manual, written by Felix von Luschan, had appeared in 1896. See also Ankermann 1914; Sarreiter 2012.

12 Doegen 1918, 7-8; "Die von der phonographischen Kommission geschaffene Lautsammlung bietet neues Material für die Sprachforschung und Auslandskunde . . . außerordentlich wertvolle wissenschaftliche Ausbeute versprechen die Aufnahmen aus den fernen und fernsten Ländern."

13 The Lautarchiv recordings were digitized between 1999 and 2007; the musical recordings at the Phonogramm-Archiv were digitized in 2013. In some cases, the same speaker/singer was recorded first by the musicologist Georg Schünemann and subsequently by the linguists. The recordings have been distributed, theoretically, according to disciplines, but actually according to the field of the researcher who recorded them. This means that there are chants, for instance, of Josef Ntwanumbi ("Twanumbee") both in the collection of the Lautarchiv (as linguistic recordings) and in the Phonogramm-Archiv (as musical recordings).

14 The original reads: ". . . in welchem Lager sind die Benin Leute? Ich habe sie nicht gesehen. In Wünsdorf ist ein Baule—Elfenbeinküste—mit 4 zugespitzten Zähnen—meines Erachtens typischer Sudannneger. Er kann fein trommeln [?]. Auch sind da gute Ful (Toucouleur) und Wolof und ein feiner Komoromann." Carl Meinhof to Felix von Luschan, July [illegible], 1915, Nachlass Felix von Luschan,

Korrespondenzen (Briefwechsel mit Carl Meinhof), Handschriftensammlung, Staatsbibliothek Berlin.

15 "Ich habe ganz und gar vergessen, mich mit Ihnen zu unterhalten über die Methode, wie man am besten an die Farbigen im Muhammedaner-Lager herankommt." Carl Meinhof to Felix von Luschan, July 16, 1915, Nachlass Felix von Luschan, Korrespondenzen (Briefwechsel mit Carl Meinhof), Handschriftensammlung, Staatsbibliothek Berlin.

16 See Lange 2011a, 2011b, 2013, 2015, 2020; see also Evans 2003; Kaplan 2013; Hilden 2021; Hoffmann 2014a, 2015a, 2018, 2020b, 2023; Hoffmann and Mnyaka 2015; and Scheer 2010.

17 On coloniality, see Quijano 2000, 2007; Qinterero and Garbe 2013.

18 Experiences of people who were subjected to anthropometric practices more than once rarely exist in writing. Yet on the recordings of Rudolf Pöch, a speaker named Xosi Tshai remarked, in 1908, that he had worked for the geologist Siegfried Passarge some years previously (Hoffmann 2020a, 88).

19 In his reports to the Academy of Science, Rudolf Pöch euphemizes his methods, while his unpublished and unfortunately also incomplete notebooks at the Naturhistorisches Museum in Vienna present another picture. See Hoffmann 2020a.

20 The Triple Entente was the coalition between the United Kingdom, France, and Tsarist Russia in World War I. For more on propaganda, see Koller 2004 and Weule 1915.

21 "Hoffentlich findet sich Gelegenheit, dass ich darauf aufmerksam mache, dass unsere linguistischen durch ihre anthropologischen Messungen ergänzt werden müssen." Carl Meinhof to Felix von Luschan, October [illegible], 1915, Nachlass Felix von Luschan, Korrespondenzen (Briefwechsel mit Carl Meinhof), Handschriftensammlung, Staatsbibliothek Berlin. On Meinhof's ideas on the connection of language and race, which he sketched in his now infamous *Hamitentheorie*, see Meinhof 1912.

22 Rudolf Pöch, notebook no. 12 (1919), p. 1125, Naturhistorisches Museum, Vienna.

23 A more detailed discussion of the connections of language and race with regard to linguistics and voice recordings would be another whole book project. Yet even if one stays with the Lautarchiv Berlin, fragmented evidence of research on race and language that may have taken place before and after the POWs of World War I were recorded appears in the form of objects. Most disturbingly, two dried human larynxes were found in a metal box some years ago, subsequently misplaced, and found again in the fall of 2017. Only recently their images were removed from the website of the collections of the Humboldt University. See Humboldt-Universität zu Berlin, Scientific Collections, Lautarchiv, "Provenance Report Regarding Two Human Larynxes in the Sound Archives at Humboldt-Universität zu Berlin," https://www.lautarchiv.hu-berlin.de/en/activities/provenienzbericht-kehlkoepfe/. The accession numbers 17 and 18, which are attached to the objects, hint at a larger collection. These two larynxes may stem from the collection of larynxes from prison camps in South West Africa

NOTES TO CHAPTER ONE

(now Namibia), which Werner Grabert had used for his research (see Grabert 1913). Further, six recordings catalogued as "Nama-Hottentottisch" at the Lautarchiv, dated 1939, which were spoken by the Polish linguist Roman Stopa, suggest connections between the Lautabteilung (now Lautarchiv) in Berlin and the linguist who described click sounds in southern African languages as "pre-articulate sounds" of primitive people. Stopa formulated his ideas in his monograph *Die Schnalze: Ihre Natur, Entwicklung und Ursprung* (1935), following Wilhelm Bleek's *On the Origins of Language* (1868). The idea of the missing link also appears in Rudolf Pöch's notebooks of his research in southern Africa (Rudolf Pöch, Notebook no. 12 (1919), p. 1125, Naturhistorisches Museum, Vienna). For an extensive discussion of German linguistics and ideas of race, see Ruth Römer (1989). And for a historical rejection of connections of language and race, see, for example, Franz Boas (1911, 1914).

24 I borrow the heading of this section from the lyrics of Tom Waits's song "The Part You Throw Away," written by Tom Waits and Kathleen Brennan, on *Blood Money* (ANTI-, 2002).

25 Carl Stumpf (1848–1936) was the founder of the Berlin Phonogramm-Archiv (1909; see Simon 2000, 13) and the chairman of the KPPK. He was a philosopher and psychologist and is seen as one of the pioneers of comparative musicology. See, for example, Stumpf 1911.

26 Letter in response to the meeting of the KPPK on May 9, 1917, signed by Carl Stumpf, Berlin Phonogramm-Archiv.

27 Mohamed Nur appears as Muhammed Nur in the publications of Maria von Tiling (later Klingenheben). Other spoken texts of POWs are published in Lange 2020.

28 The distinction between *langue* (French for language) and *parole* (speech) was introduced with Ferdinand de Saussure's *Course in General Linguistics* (1916), which described the two concepts as systematic principles of language. This distinction may now belong to the realm of theoretical fiction in linguistics. While *langue* was seen as the body of abstract systematic rules of a language as a signifying system, and therefore also the nonpersonal precondition for any meaningful utterance, *parole* was described as the individual, performed, concrete utterance. The KPPK could practically record only concrete utterances of the prisoners. These came in dialects and in various registers of speech and were gendered and positioned. Even with regard to their interest in phonetics, the KPPK members were far from recording what they may have understood as "general" (in absence of a standardized) pronunciation of the languages. This becomes clear with the acoustic documents, for instance, of Abdoulaye Niang's strong accent and dialect, but also with Carl Meinhof's comment on the pronunciation of the clicks by the isiXhosa speaker Josef Ntwanumbi ("Twanumbee") (see Fragment IV). The fact that this speaker may not have spoken his first language for many years is not taken into account.

29 The portrait also appears as a digital, pixelated screenprint in the exhibition catalogue *Deutscher Kolonialismus: Fragmente seiner Geschichte und Gegenwart* (2016,

15), which again has images of recording, of records, and of a personal card created for the musical recordings, but no transcribed text of a recording.

30 The *Linguistic Survey of India* (1903–28), compiled and edited by George Abraham Grierson, is a monumental language survey, describing 179 languages and over 500 dialects (a distinction that is in itself questionable).

31 Some of the written files indicate *Kriegsgefangenenlager* as the standard location of recording, while others do not. After all, the Prussian scheme of registration was not as systematic as reputed.

32 The publications about the recordings with POWs issued by the Lautbibliothek do not mention the camps or the Kommission, although the recordings are based on its operation. Maria von Tiling does not mention that Mohamed Nur, with whom she worked over a long period, had been released from Ruhleben on Meinhof's request, to become a language assistant in Hamburg (see von Tiling 1918/19, 1925). Carl Meinhof and Martin Heepe do mention the POW camps. In an article published in 1939, Meinhof writes that he had spoken with "an educated Bira for several days" (1939, 147) in a POW camp in Münster. He does not speak of a systematic survey, nor does he mention acoustic recordings. Martin Heepe's (1920) dissertation is based on his research into languages from Comoros, which he recorded during the war. In the foreword of his book on Comoro languages, he mentions his work with the Kommission.

33 For a recent discussion of the position of the "native informant," see Phadi and Pakade 2016.

34 Several selective processes antedate the actual archiving of examples of African languages that are held as records in the Lautarchiv. For a discussion of racial preconceptions in the French military (before and during World War I) about specific *races guerrieres*, which played a role in the selection of areas and ethnic groups from which soldiers were recruited, see Lunn 1990. The number of recordings produced in specific languages does not mirror the demographic situation in the camp, but speaks of the interests of particular researchers. Martin Heepe, who produced most of the recordings in African languages, had a keen interest in languages from Madagascar and Comoros. On the basis of these recordings, he wrote *Die Komorendialekte Ngazidja, Nzwani und Mwali*, which was published in 1920 by the Seminar für Kolonialsprachen based on his doctoral dissertation.

35 The linguistic recordings that are now held at the Lautarchiv were recorded with a gramophone (on records); the recordings of music for the Phonogramm-Archiv were produced with a phonograph on wax cylinders.

36 In the Lautarchiv he appears as "Buru" with no other details given but that he was a "Baule Mann" who did not speak any language but "Baule." In the publication of Josef Weninger, a "Goli Bru" appears as number 3870, again as "a Baule" speaking no other language than "Baule" (Baoulé). Weninger states that the speaker came from Ivory Coast, was married, and had one child. How he communicated with the prisoner he does not mention.

37 Transcribed and translated by Serigne Matar Niang.

38 On concepts of hearing and listening in West Africa, see Baumann 1997 and Chernoff 1997.

39 Fritz Bose (1906–1975) was a musicologist and head of the musicological division of the Forschungsgemeinschaft Deutsches Ahnenerbe, a research institute founded by Heinrich Himmler with the aim of bolstering the racial ideology of the National Socialist (NS) state. The institute was directly connected to the SS. After the war, Bose worked at the Staatliches Institut für Musikforschung in Berlin (Bundesrepublik Deutschland).

40 Egon von Eickstedt (1892–1965) was a student of Felix von Luschan. He measured 1,784 prisoners in sixteen camps during World War I. His doctoral dissertation was built on anthropometric data from the camps, which was partially published as "Rassenelemente der Sikhs" in the *Zeitschrift für Ethnologie* in 1921. Although he lectured on "racial elements," developed theories on the races of the world and their hierarchies, worked with Eugen Fischer, and contributed as a consultant in surveying the "racial origin" of people in the NS state, he was able to continue working as an anthropologist in postwar Germany at the universities of Leipzig and Mainz.

41 These terms describe fascist fantasies of race: "Indians of the forest race" (from the Latin *silva* for "forest"), "Negroes of the Sudanic or paleo-negroid races," "Europeans of the Nordic races." It is just one of the odd footnotes of the history of race studies that in Josef Weninger's research, Abdoulaye Niang became a "Hamitic Negro."

42 See also Radano and von Bohlmann 2000; Sterne 2008.

43 Carl Meinhof to Felix von Luschan, December 13, 1917, Nachlass Felix von Luschan, Korrespondenzen (Briefwechsel mit Carl Meinhof), Handschriftensammlung, Staatsbibliothek Berlin.

44 The Sultan of the Ottoman Empire had called a jihad against the Triple Entente, which was the basis for the Germans' attempt to convince Muslim soldiers to join the army of the Central Powers. See Lange 2011b, 103; and Roy, Liebau, and Ahuja 2011.

45 Carl Meinhof to Felix von Luschan, October [illegible], 1915:

> Ich wollte Ihnen noch mitteilen, dass sich in Wünsdorf ein Mossi aus der Nähe von Fada n Gourma befindet. Soviel ich sehe, ist der Mann anthropologisch sehr interessant, sein ganzes Gesicht ist mit langen Narben bedeckt in einer Weise, die ich noch nicht gesehen habe. Der Gesichtsausdruck und die sehr helle Farbe erinnern mich unwillkürlich an Buschleute, obwohl der Mann durchaus nicht klein ist. Die Nase ist an der Wurzel vollständig eingedrückt und die Augen liegen außerordentlich tief, er macht die Augen auch wenig auf. Der gesamte Eindruck ist für den Unkundigen geradezu abschreckend und doch ist es ein recht gescheuter Mensch und als ich mir Texte von ihm geben ließ, redete er besonders davon, dass er seinen Vater und seine Mutter, seine Frau und seine Kinder, deren er zwei hat, so lange nicht gesehen hätte und befürchtet, er würde sterben, ehe

er sie wieder sehe. Es ist nur der eine Mossi da, Sie würden ihn also bald herausfinden.

I also wanted to report that in Wünsdorf there is a Mossi from Fada n Gourma. As far as I can tell, the man is of anthropological interest, his entire face is covered with long scars in a way I haven't seen before. The facial expression and the very fair color reminds me of bushmen, although the man is not short. The root of the nose is caved in and his eyes are cavernous, he opens them barely. The overall impression is downright frightening for the uninformed (*den Unkundigen*), yet he is smart and when I asked for texts he spoke of having not seen his father and mother, his wife and children, of which he has two. He fears he will die before he sees them again. There is only one Mossi, you would find him easily.

46 From 1917 onward, about three thousand African and Indian prisoners were sent to Romania, in the hope, as Gerhard Höpp (1997, 50–51) writes, that this would decrease the alarming number of deaths, especially from tuberculosis, in Wünsdorf and Zossen. In a letter to his wife, written in March 1917, Leo Frobenius claimed that cases of illness had massively decreased (Kuba 2015, 109–10).

47 I thank Katarina Matiasek for searching through the files and patiently explaining the logic of presences and absences in this archive to me.

48 Not all methods of anthropometric documentation were applied to every specific case. Of Farnana, however, even a plaster cast was made.

49 On stereo photographs, see Matiasek 2017; Lombard 2018.

50 The Frobenius Institut in Frankfurt holds the collection of Leo Frobenius, which includes the photographs from the camps in Romania.

51 The translation given here is an assemblage of the translations by Serigne Matar Niang and by Fatou Cissé Kane, who lives in Cologne. The translation of this particular recording was very difficult: each of the translators interpreted words differently, and the lyrics appear to entail some obscenity. Because I do not know Wolof, I am unable to say more about it at this point.

CHAPTER 2. MOHAMED NUR: TRACES IN ARCHIVES, LINGUISTIC TEXTS, AND MUSEUMS IN GERMANY

1 In Ruhleben, recordings were made with internees from Nigeria, Sierra Leone, Malaysia, Samoa, Tahiti, and the United Kingdom.

2 The names of those recorded were often misspelled by the KPPK and continue to appear this way in the Lautarchiv and in other archives. It is impossible to restore them retrospectively. This book uses spellings based on naming and spelling practices in the men's respective languages and cultures. Mohamed Nur appears as Muhammed Nur in Maria von Tiling's work and as Mohamed Nor in the files of the Red Cross. Josef Ntwanumbi appears as Josef Twanumbee in the Lautarchiv, which is simply an impossible misspelling.

3   See my sound installation *Foreign Subjects* produced for and with the Bergen Assembly 2019 (an art triennale in Norway), which presented spoken and sung texts by Josef Ntwanumbi and Mohamed Nur, both of whom were interned in Ruhleben. https://2019.bergenassembly.no/contributors/anette-hoffmann/.

4   See the index to this list: *Index of British Fishermen and Merchant Seamen Taken Prisoner of War 1914–1918*, https://www.spw-surrey.com/MT9/POW2012.aspx?name=*. The original document is the *List of Mercant Seamen and Fishermen Detained as Prisoners of War in Germany, Austria-Hungary and Turkey* (London: Board of Trade, 1918), National Archives, Kew MT 9/1238.

5   Albert Grohs (sometimes written Groß; 1880–1935) was a German documentary and press photographer. He ran a stock photo agency (*Bildagentur*) in Berlin from 1908 onward, and he collaborated with the Berliner Illustrationsgesellschaft. His photo archive was lost during World War II. According to the archive of the Harvard Law School Library, which holds the Maurice L. Ettinghausen Collection, the photograph shown in plate 4 was acquired by Ettinghausen in the 1930s. Ettinghausen was himself interned in Ruhleben and later collected material on Ruhleben. The German caption (verso) reads: "Alle Rechte vorbehalten. A. Grohs Illustrationsverlag, Berlin, Ammerstr 48." The English caption reads: "Black internees in their barrack. Grohs, A., photographer." The photos by Grohs in the Ettinghausen collection at Harvard are listed as "40 Propaganda Photos."

6   Auguststrasse is not far from where I lived, in Berlin Mitte. Where number 4 must have been there is now a playground, the layout of which is quite typical in gaps that resulted from the bombing of Berlin during World War II.

7   On the long history and pertinent presence of the relationship between researchers and "native informants," see Phadi and Pakade 2016. On colonial linguistics, see Errington 2008; Irvine 2008; Deumert, Storch, and Shepherd 2020, among others.

8   After 1919, this entity became the Seminar für Eingeborenensprachen (Seminar for Native Languages).

9   Carl Meinhof to Felix von Luschan, December 1, 1917, Nachlass Felix von Luschan, Korrespondenzen (Briefwechsel mit Carl Meinhof), Handschriftensammlung, Staatsbibliothek Berlin.

10 Sarah Pugach writes that Maria von Tiling was involved in the recordings with POWs, but this does not appear in any documentation at the Lautarchiv (Pugach 2012, 161).

11 Carl Meinhof to Oberschulbehörde, March 21, 1921, Personalakte Muhammed Nur, 361-6-V-180/41, Staatsarchiv Hamburg.

12 On the status and situation of the *Lehrgehilfen* (the African language assistants), see, for instance, Stoecker 2013.

13 See von Tiling 1925, 43–92.

14 The exhibition was at the Wallach Art Gallery, Columbia University, New York, October 2018 to February 2019, and the Musée d'Orsay in Paris from March to July 2019. For the catalogue, see Murrell 2018. Two paintings of Mohamed Nur by

Max Slevogt (one of which names the sitter as Hassanó) are held by the Museum Kunstpalast in Düsseldorf, Germany (see plate 10). A drawing by Slevogt with Nur unnamed is held by the Art Institute of Chicago; see Max Slevogt, "Somali Warrior" (1912), drypoint on cream wove paper, Art Institute of Chicago, 1975.1041, https://www.artic.edu/artworks/49850/somali-warrior.

15 Rudolf Jacob Zeller (1880–1948) was a Jewish artist who made a living in Hamburg mainly by painting portraits and teaching art. He emigrated to the Netherlands in 1938.

16 As a model for Max Slevogt, Nur may have called himself "Hassanó," which is the name Slevogt mentions in the title of one painting (see plate 10, top right). For the full painting, see Max Slevogt, *Brustbildnis des Somali Hassanó*, accessed June 9, 2023, https://www.bildergipfel.de/kunstdrucke/kunststile_und_epochen/impressionismus/brustbildnis_des_somali_hassano_1912_max_slevogt.

Ludwig Gerhardt (2004, 442) misidentifies the painting shown in plate 6 in reproduction as a depiction of Juma bin Abdallah. Alexander Meckelburg (Hiob Ludolf Center), with whom I first spoke in October 2016, told me that nothing was known about the image at the Asien-Afrika-Institut, which held the painting at that point. In the meantime, the painting has been restored and was part of an exhibition on the history of the University of Hamburg, which opened in 2019; see https://www.jubilaeum.uni-hamburg.de/100-jahre-uni/ausstellung-hauptgebaeude.html.

17 Another invasive practice the interned men were subjected to was production of casts of the mouth: Wilhelm Doegen's brother, a dentist, produced *Palatogramme* (casts of the tongue and the mouth) in the internment camps to "study rare sounds of exotic tribes at the very site of their origin"(Doegen 1925, 16; "um seltene Laute exotischer Völkerstämme am Orte ihrer Enstehung zu studieren"). A longer discussion on language and race in relation to sound recordings is needed, but cannot be offered here. Meinhof's writings, both published and unpublished, demonstrate his ideas in this respect (see, for instance, Meinhof 1913, 6). Werner Grabert's (1913) comparative research on human larynxes of African and European origin, and of monkeys, is relevant here. Another gruesome (re)discovery was the unlabeled box containing two dried human larynges, found in the Berlin Lautarchiv some years ago (see above pp. 165–66, note 23). The larynxes are part of the collections of the Humboldt University of Berlin. Whether these were related to research on language and race, I have not been able to confirm.

18 Britta Lange, Institut für Kulturwissenschaft Humboldt Universität Berlin, personal communication with the author, August 2018. On the photographs of African POWs, see Doegen 1925, 33.

19 On re-assembling archival objects, see Harrison, Byrne, and Clarke 2013.

20 On the history of auscultation as an audile technology of diagnosis, see Sterne 2003.

21 Newspaper advertisement in the daily *Hamburger Anzeiger*, May 21, 1927, cited in Zedelmaier 2007, 183.

22 "Treptow: Die Deutsche Colonial-Ausstellung von 1896 im Treptower Park," Afrika in Berlin: Ein Stadtspaziergang des Deutschen Historischen Museums, accessed June 9, 2023, https://www.dhm.de/archiv/ausstellungen/namibia/stadtspaziergang/treptow.htm.

23 Abraham Ulrikab, from Labrador, arrived in Hamburg with his wife and three children to work for Hagenbeck in 1880. The entire family died of smallpox in hospitals in Paris and Krefeld in 1881. Instead of being buried, the human remains of Abraham Ulrikab and his family ended up in museum collections in Paris and Berlin. Ulrikab's original diary is missing, therefore the publication is based on a later translation from Inuktitut (Ulrikab et al. 2005). Friedrich (or Frederik) Maharero was the son of Paramount Chief Samuel Maharero. He was twenty-two years old when he arrived in Germany. His aim in making the journey was to gain information and to establish diplomatic connections in the country (Bechhaus-Gerst 2003, 102). In September 1896 he was granted an audience with Kaiser Wilhelm II, together with Petrus Witbooi and Josaphat Kamatoto. Samuel Maharero lived in exile in Botswana after the colonial war in German South West Africa (1904–8), but was eventually able to return to Okahandja in 1932. About his stay in Berlin he wrote: "I was shown to the Kaiser who did not know his black subjects. We were there for a year. We were not taught anything, we just rode on horses and we dressed and were drilled like soldiers" (Bechhaus-Gerst 2003, 103, as cited in Katjavivi 1988, 11).

24 See, for instance, the Clemens Radauer collection of postcards issued by the shows: "Collection Radauer," Human Zoos, accessed June 9, 2023, https://humanzoos.net/?page_id=4315..

25 Letter from E. Holz to the Phonogramm-Archiv, Berlin, undated; Phonogramm-Archiv, Berlin.

26 The recordings are published in the CD set *Black Europe: The Sounds and Images of Black People in Europe—Pre 1927*, Bear Family Records, 2012, https://black-europe.com/the-black-europe-box-set/.

27 Recording No. 1, "Archiv Somali," is held at the Berlin Phonogramm-Archiv. After this first part, the song turns into a provocation. First an unspecified person is verbally abused, and then women in Berlin are berated and insulted. It is difficult to discern the meaning of these verses: they could be the verbal part of the staged battles (*diradire*), which members of the ensemble had to perform. They could also be what Said Samatar (1982, 74) calls a "diatribe," uttered to disgrace an enemy, which is especially effective if this person cannot counter the verbal attack with verse, for instance, because he or she does not understand it. The recording is part of my sound installation *Foreign Subjects*, which was shown at the Bergen Assembly in 2019.

28 See also Das 2011a; Lange 2011b, 93; Stoecker 2008.

29 See, for instance, Johler, Marchetti, and Scheer 2010; Lange 2013; Berner 2003, 2004; Evans 2010a, 2010b.

30 Käthe Kollwitz, diary entry, September 29, 1914, as cited in Burkard and Lebret 2015, 147.
31 For the use of the term *Völkerschau* or *Völkerzirkus* in reference to non-European soldiers of World War I, see Frobenius 1916; Stiehl 1916; and Weule 1915, among others.
32 From the pamphlet on the Somali Dorf, printed by Friedrich Kaiser in Vienna. A copy is held by the archive of the Department for Evolutionary Anthropology, University of Vienna.
33 Another set of recordings with Nur held at the Berlin Phonogramm-Archiv, recorded by the musicologist Georg Schünemann, probably earlier, is undated.
34 All recordings of the World War I POWs in the Lautarchiv were digitized between 1999 and 2007.
35 Bodhari Warsame comments that here Nur moves from chanting to speaking angrily in a simulated or staged dialogue.
36 Translated and transcribed by Bodhari Warsame.
37 *Geerar* is a subgenre of *maanso* (Banti and Giannattasio 1996, 98).
38 Transcription and translation by Bodhari Warsame.
39 Nur names Shihiri, but not the man who was probably his teacher; the latter appears in Samatar; he came from Berbera, like Nur (Samatar 1982, 127).
40 Nur's account is consistent with Samatar's reading of the historical events.
41 Dates were often used as provisions during travels.
42 Not all the known paintings and drawings can be located: the Museum Kunstpalast in Düsseldorf, Germany, holds the painting *Der Sieger/Kriegsbeute* (The Victor/Spoils of war) and *Brustbild des Hassanó* (Portrait of Hassanó), which I was able to see in 2019 (plate 10). For my exhibition *Der Krieg und die Grammatik* at the MARKK in Hamburg, Jannik Franzen and I were allowed to film these artworks from the Kunstpalast. A depiction of Nur squatting titled *Hockender Aschanti/Neger mit vorgestreckten Armen* (Squatting Ashanti/Negro with arms outstretched) is mentioned as a missing artwork (lost in 1938) with the ID 388691, at Lost Art–Datenbank: Deutsches Zentrum Kulturgutverluste, accessed January 2019, https://www.lostart.de/de/start. Another image of him squatting, entitled *Hockender Afrikaner* (Squatting African), was sold by the Viennese auction house Dorotheum to a private collector in Germany some years ago. Also sold by the same auction house in November 2017 was a painting in which Nur is *Der Tänzer* (The dancer). The Art Institute in Chicago holds a sketch with the title "Somali Warrior" (see note 14 above). Another similar sketch, drawn on the back of an envelope, is held by a private collector, who also owns the *Hockender Afrikaner*. And a painting of Nur with spear and feathers appears in the journal *Deutsche Kunst und Dekoration* 28 (September 1915): 418.
43 For a description, see Guthmann 1920, 157. The German term *Herrenmensch* relates to racist hierarchies. In this case, combined with *dunkelhäutiger* (dark-skinned), it refers to tall, slender people in Africa, who were seen as superior to others and therefore entitled to dominate others. The term *Herrenmensch*

became prominent in the ideology of the Nazis, who staged themselves as such (Guthmann 1920).

CHAPTER 3. ALBERT KUDJABO AND STEPHAN BISCHOFF: MYSTERIOUS SOUNDS, OPAQUE LANGUAGES, AND OTHERWORLDLY VOICES

1   See Brosens 2013, 94. Griet Brosens writes about thirty-two Congolese and black Belgian soldiers. She believes that Albert Kudjabo came to Ghent, Belgium, as a servant, but I have not seen any archival trace of this.
2   This particular drum could not be found when I produced the first part of my exhibition *Der Krieg und die Grammatik* at the MARKK in Hamburg (formerly Museum für Völkerkunde, founded in 1879) in fall 2019; see https://anettehoffmann.com/der-krieg-und-die-grammatik-i-2019/.
3   As with Abdoulaye Niang's archival traces discussed in chapter 1, the acoustic and written files that document Albert Kudjabo's presence in the camp point to a network of researchers and their interests. In this case, these interests revolve around drum language, or "language surrogates" (see Sebeok and Umiker Sebeok 1976). Notably, a letter from the German physician Albert Schweitzer, in which he expressed concern about the disappearance of rowing songs in Gabon, added rowing songs to the shopping list of the linguists (Ames 2003; Simon 2000). This example of the autopoesis of the archive is more than just an episode of colonial knowledge production, because it speaks of the productivity of the will to know and the prisoners' performances as responses to requests that are often not found in the documentation of the archive itself (see Hoffmann 2020b).

   The desire to record rowing songs, which I have discussed elsewhere (Hoffmann 2020b), is an example of the circular moves of archive: the recordings' existence is directly related to the letter from Schweitzer to Carl Stumpf, written in Gabon in 1914. Schweitzer articulates his concern about the disappearance of rowing songs with the introduction of motorboats. As a result of the letter to Stumpf, who was director of the Berlin Phonogramm-Archiv, eight rowing songs were recorded with Albert Kudjabo and Asmani ben Ahmad from Comoros. To my knowledge, they were never translated nor were they published in any form.
4   See, for instance, Renton, Seddon, and Zeilig 2007; and Hochschild 1999. The Congo Free State became the Belgian Congo in 1908.
5   See Dossier 1.136.525, Archives Generals, Brussels; Brosens 2013; and "Le corps des vontaires au Congo," *Mbokamosika* (blog), August 24, 2012, https://www.mbokamosika.com/2014/03/le-corps-des-volontaires-congolais-1914-1918.html. Griet Brosens states that Kudjabo arrived in Belgium as a servant (she writes "boy"), but I have not seen evidence for this in any of the archival records (Brosens 2013, 40).
6   See Renton, Seddon, and Zeilig 2007, 25.

7   Transcription and translation by Gilbert Katanabo Muhito and Faustin Sambu Avetsu.
8   An example is the work of the Austrian anthropologists Paul Schebesta and, later, Martin Gusinde, who were interested in people they described as "pygmies." In his bewildering prose Schebesta (1934, 129) manages to attribute complete, ahistorical stasis to the people who inhabited the area of Ituri, while at the same time speaking of the forced relocation and indentured labor of the people who were building roads. By the time of his arrival in the area, the inhabitants had experienced the impact of slave trade, Belgian colonization, and the massive transformation of land use from subsistence farming to plantations and mines. In Schebesta's tale, drums appear in close proximity to cannibalism, of which he accuses the "Bira of the forest" (*Waldbira*). Language appears as a marker for ethnic connections among the population of Ituri as well as for offering clues for the understanding of migration patterns. Schebesta describes the contrast of affluence in an area where agriculture seemed easy to the aftermath of colonial relocation, which appears as a (normalized) transition into a capitalist present. Such racializing anthropometric research was published as late as 1975 (Sporcq 1975).
9   For this interpretation I am indebted to the participants of one of the sound studies workshops I organized at the Archive and Public Culture Research Initiative in Cape Town in 2012. During these workshops we listened to some work songs and discussed the expressive possibilities of speaking with rhythm. Special thanks for this discussion go to radio maker Anthony Sloane and musician and musicologist Thokozani Mhlambi.
10 The transcription was first published in the journal *Vox*: "Die Trommelsprache in Afrika und in der Südsee" (Drum language in Africa and the South Pacific), Hamburger phonetische Vorträge 5, *Vox*, nos. 4–5 (1916): 179–208.
11 By March 1917, when the recording took place, Albert Kudjabo had been transferred to Münster.
12 Carl Meinhof to Felix von Luschan, November 1, 1916, Nachlass Felix von Luschan, Korrespondenzen (Briefwechsel mit Carl Meinhof), Handschriftensammlung, Staatsbibliothek Berlin. The original reads as follows:

> In dem Gefangenenlager zu Soltau in Hannover fand ich einen Eingeborenen aus dem Kongostaat, der Trommelsprache kann. Ich will versuchen, mit dem Mann einige Aufnahmen zu machen, und brauche dazu drei Holztrommeln verschiedener Größe. Herr Professor Thilenius will die Freundlichkeit haben, mir diese Trommeln zu besorgen und würde auch eventuell bereit sein, mit mir nach Soltau hinüber zu fahren, um mir bei den Aufnahmen zu helfen. Damit würden wir dann aber in das ethnographische Gebiet hinübergreifen und ich will das nicht tun, ohne Sie um Ihre Zustimmung zu bitten. Es scheint, dass in dieser Sprache, wie im Tschi, die Tonhöhen der gesprochenen Sprache mitgetrommelt werden und dieses linguistische Problem ist es, was ich eigentlich verfolge. Natürlich werde ich versuchen, bei dieser Gelegenheit so viel Trommelaufnahmen zu bekommen, wie möglich ist. Ich

erbitte, Ihre freundliche Zustimmung zu diesem Plan oder, wenn Sie Bedenken haben, um Mitteilung dieser Bedenken. Mir liegt daran, auch in dieser Sache, wie sonst immer, Hand in Hand mit Ihnen zu arbeiten.

See also the first note in Thilenius, Meinhof, and Heinitz [1916] 1976.

13 For a discussion of the role of the Phonetic Laboratory in research on African languages, see Pugach 2018.
14 "Speaking drums" were apparently sought after for museum collections. The Royal Africa Museum in Tervuren, Belgium, lists 421 slit drums from Congo.
15 Thilenius, Meinhof, and Heinitz [1916] 1976, 1:7–8. The original reads:

> Ich hatte zunächst Gelegenheit die Trommelsprache der Duala in Kamerun kennenzulernen. Das Ergebnis einer langen Beschäftigung mit dem Gegenstand war, dass die Duala imstande sind, durch eine große Anzahl von Signalen, die sie mithilfe der Holztrommeln geben können, sich über alle möglichen Dinge zu verständigen. Nachrichten über durchziehende Expeditionen können auf diese Weise weitergegeben werden. Man kann einen mißliebigen Gegner aus der Ferne ungestraft mit der Hilfe dieser Signale schmähen und ist in der Lage, da die Signale sehr weit zu hören sind und von anderen Trommelkünstlern weitergegeben werden, jede beliebige Nachricht mit erstaunlicher Schnelligkeit zu verbreiten. Es gehört ein ziemlich langes Studium dazu, um diese Kunst zu lernen, und der gewöhnliche Mann versteht in der Regel nur wenige Signale, wie z.B. das Signal für Krieg, komm her und vor allen Dingen seinen eigenen Namen. Sogar Hunde sollen darauf dressiert werden, zu kommen, wenn ihr Name getrommelt wird. . . . Auf welchen Prinzip die Trommelsprache der Duala aufgebaut ist, ist mir noch heute nicht bekannt.

This is an excerpt of a transcription of Meinhof's contribution to the phonetic evening, which is quoted in Thilenius, Meinhof and Heinitz [1916] 1976, 7–8.

16 The fact that drum language was identified as a threat led, for instance, to the 1740 prohibition against enslaved Africans owning drums in the United States. The Slave Code of South Carolina described it as "absolutely necessary to the safety of this Province, that all due care be taken to restrain Negroes from using or keeping of drums, which may call together, or give sign notice to one another of their wicked designs and purposes" (Article 36, 1740); see https://ushistoryscene.com/article/excerpts-south-carolina-slave-code-1740-no-670-1749//.
17 In Western academic publications, the consideration of or interest in the aesthetics and literary quality of drum language seems to feature much later, notably in the 1960s (concurrent with or after the independence of many African states) in the work of the Ghanaian musicologist Joseph Hanson Kwabena Nketia (1963).
18 *Berliner Illustrirte Wochenschau* 9 (33), March 2, 1924.

19   See, for instance, *Sanders of the River* (1935) and its satire, *Steve of the River* (1937); *Tarzan the Ape Man* (1932); and also *The Nun's Story* (1959), which partially takes place (and was produced) in the Congo, and in which drum language is also appropriated by the mission to call children to school.
20   From Wilhelm Heinitz, "Reise nach Südfrankreich zur weiteren Erforschung der Trommelsprache," unpublished manuscript, as cited in Stoecker 2008, 142.
21   In his introduction to the second volume of *Speech Surrogates: Drum and Whistle Systems*, Thomas Sebeok notes that in the early years of research on speech surrogates (1887–1907), publications on African language surrogates (mostly drum language) accounted for 71 percent of all publications (Sebeok and Umiker-Sebeok, 1976, 2:xxii).
22   German radio broadcasting began in December 1923.
23   From the recording of the broadcast *Funkstunde*, with the title "Stimmen der Völker," May 30, 1924, NDR Radio Archive, Germany.
24   The original German follows (transcribed): "Er schafft alte Sprachen, alte Musiken, seltenen Musik, und Worte und Laute, die wir sonst nicht zu hören bekommen." "Stimmen der Völker," NDR Radio Archive.
25   "Stimmen der Völker," NDR Radio Archive. There must have been photographs of the African POWs as well, but these are no longer at the Lautarchiv.
26   Carl Stumpf (1848–1936) was *Tonpsychologe*, philosopher, and founder of the Berlin Phonogramm-Archiv in 1900. In 1915, when the Kommission began the recordings, Erich Moritz von Hornbostel was the director of the Phonogramm-Archiv (see von Hornbostel [1905] 1986, 1928; von Hornbostel and Abraham 1904).
27   See "Introduction," Lautarchiv Humboldt-Universität zu Berlin, accessed June 13, 2023, https://www.lautarchiv.hu-berlin.de/. See also Lange 2011a, 2013; Kaplan 2013; Hoffmann 2014a, 2015a; and Hoffmann and Mnyaka 2015.
28   *Berliner Illustrierte Wochenschau* 9 (33), 1924.
29   Stoler's notion of the "rot that remains" is based on Derek Walcott's poem "Ruins of a Great House" (Walcott 1990, 19–20).
30   See, for instance, Mbembe 2014.
31   On the coloniality of the denial of the intersubjective character of knowledge production, see Quijano 2007.
32   The statute of the Union congolaise is filed in W. E. B. Du Bois's papers at the University of Massachusetts, Amherst, which also holds Paul Panda Farnana's letters to Du Bois. See also Kongolo 2011.
33   Odette Kudjabo, personal communication with the author, August 2018.
34   German Togoland was the name of a colony under German rule, encompassing what is now Togo and parts of present-day Ghana, between 1884 and 1914.
35   On nationalist ideas of the Norddeutsche Missionsgesellschaft (Northern German Missionary Society) in Togoland, see Meyer 2002. See also Zahn 1895.
36   Stephan Bischoff and Josef Ntwanumbi may be in the photograph taken in Ruhleben that shows Mohamed Nur, Carl Meinhof, and Wilhelm Doegen, together with three other African internees whom I cannot identify (plate 5).

37 Poetic interpretations of language examples and sound recordings are often problematic because they are extracted from archives with little (or no) consideration of the archival background and circumstances of their production. A prominent example is the archive of the linguists Wilhelm Bleek and Lucy Lloyd, which has become a source of inspiration for literature, theater, and art (see The Digital Bleek and Lloyd, accessed June 13, 2023, http://lloydbleekcollection.cs.uct.ac.za/). Texts revered as authentic traces of "lost" culture and languages are often, in form and content, shaped by specific recording situations. In particular, the repetition of words served to enable the linguist to transcribe the spoken texts. As with the production of sound recordings, questions or situations which prompted the utterance of the words and sentences that entered the archive were often not documented. For a critique of the uses of the Bleek and Lloyd Collection, see Rassool 2006.

38 On reading this chapter, Anne Storch remarked that these anecdotes and, with them, the questions of who hears or can perceive what still circulate in the field of *Afrikanistik* (African studies); personal communication with the author, 2020. At the same time, she pointed out that the example of the congregation members as piglets may present another racist subtext, referring to those who need the mission to become civilized.

39 Victor Toso (1893–1916) had been employed as a language assistant from 1912 onward. He succumbed to a lung disease at the hospital St. Georg in Hamburg. See Meyer-Bahlburg and Wolff 1986.

40 Transcription and translation by Johannes Ossey.

41 The information on these recordings is contradictory: Johannes Ossey hears two songs in a codified language; the Lautarchiv registers one and describes another one as a war song with drums.

42 Afelevo's report of 1894 was later republished by Afelevo and Westermann in 1930.

43 *Muttersprache* (mother tongue) is not a very useful concept for people who grow up bi- or polylingual and whose family structures and child care concepts do not necessarily fit into European models of the nuclear family.

44 In the regional slave trade, being sold as a child did not necessarily mean lifelong slavery, nor were offspring of enslaved people automatically born into chattel status.

45 For a discussion of slave trade among the Anlo-Ewe, the entangled history of Christianity and Eweness, and the status of slave children, see Meera Venkatachalam, 2012.

46 Andrew Zimmermann (2015, 269) writes: "The Yewe secret society facilitated resistance to German colonialism in Tove. Yewe referred not only to the *trowo* (gods/fetishes) but rather to the secrecy of the society itself."

47 See Weidman 2015; Dolar 2006, 2007; Kolesch and Krämer 2006; Cavarero 2012.

48 The German linguist Diedrich Westermann also wrote about Yevegbe in 1902 and 1930.

49 On secret languages, see, for instance, Arich-Gerz and Effing (2017) and Heller-Roazen (2013).
50 Personalakte of Stephan Bischoff, 361-6-V-158, Staatsarchiv Hamburg.
51 National Archives, Kew, MT 9/1094.
52 Carl Meinhof to Oberschulbehörder (Board of University), May 11, 1917; Personalakte of Stephan Bischoff, 361-6-V-158, Staatsarchiv Hamburg. The original German follows:

> Die Oberschulbehörde beehre ich mich zu bitten, mir die Beschäftigung eines Sprachgehilfen für Ewe an Stelle des verstorbenen Victor Toso zu gestatten. Seit 1,5 Jahre habe ich den Unterricht um Ewe ohne Mitwirkung eines Sprachgehilfen erteilen müssen, worunter die Ausbildung der Zuhörer natürlich empfindlich leidet. Ausserdem habe ich die Mitarbeit des Sprachgehilfen bei der Drucklegung von Ewe-texten, z. B. in Heft 1 des laufenden Jahrgangs der Zeitschrift für Kolonialsprachen und bei den Übersetzungen von Ewe Texten, wie sie mir nich vorliegen, empfindlich vermisst. Mein Bemühungen einen geeigneteren Ersatz zu finden, waren bisher vergeblich, ich habe nunmehr aber im Engländerlager in Ruhleben einen geeigneten Mann gefunden. Stephan Bischoff ist geboren in Keta an der englischen Goldküste und spricht Ewe als sein Muttersprache. Sein Vater war Evangelist der Bremer Mission und der St. Bischoff spricht gut Deutsch und macht den Eindruck eines verständigen und unterrichteten Mannes. Er war Lademeister auf einem Dampfschiff und wurde als britischer Untertan am Kriegsbeginn interniert. Ich halte es für möglich. Dass man ihn entweder entlassen oder beurlauben wird und bitte ganz ergebenst, einen dahin gehenden Antrag and die Kommandatur des Engländerlagers in Ruhleben bei Spandau zu richten. Ich schlage vor ihm ebenso wie die Sprachgehilfen Messi und Makembe mit monatlich 130 Mark, ausschließlich der Kriegsbeihilfe zu besolden und ihn einstweilen nur für das laufende Semester zu beschäftigen. Für den Unterricht haben sich mehrere Lehrerinnen der Norddeutschen Mission gemeldet, die aber nur teilnehmen wollen, wenn ein Sprachgehilfe da ist, da sie selbst in Togo gewesen sind und die Elemente der Sprache beherrschen. Ohne einen Sprachgehilfen wäre ich nicht imstande, einen für diese Zuhörerinnen förderlichen Unterricht zu erteilen. Für den Anfängerkurs in Ewe hat sich niemand gemeldet. Es wird für die Erteilung von Auskünften über die Verhältnisse in Togo und zur Übersetzung von etwa einlaufenden Briefen die Mitarbeit des Sprachgehilfen von großem Nutzen sein.

53 On the situation of language assistants in Hamburg, see also Pugach 2012; Stoecker 2008.
54 Personalakte of Stephan Bischoff, 361-6-V-158, Staatsarchiv Hamburg.
55 Personalakte of Stephan Bischoff, 361-6-V-158, Staatsarchiv Hamburg.
56 On positionalities in orature and the license to criticize, see Hoffmann 2015b and 2012.

57 Adjaï Paulin Oloukpona-Yinnon (1996, 200) writes that von Rentzell came to Togo as an officer in the German colonial troops at the age of twenty-four.

58 On acoustemology, see Feld 2015. For a discussion on colonial acoustemology of perceived radical difference, see also Ochoa Gautier 2014.

59 Correspondence between Hans Gruner in Krachi and Felix von Luschan in Berlin makes clear that von Luschan was informed about the confiscation of the so-called fetish objects immediately after the desecration of the shrine and articulated keen interest in them. Hans Gruner to Felix von Luschan, November 23, 1896, E 1386/1896, Ethnologisches Museum, Staatliche Museen zu Berlin. https://id.smb.museum/object/788129/e-1386-1896

60 For a more detailed interpretation of the role of Hausa traders in this conflict, see Apoh 2020 and Maier 1980.

61 Hüsgen 2020, 11; Hans Gruner to Felix von Luschan, November 23, 1896, E 1386/1896, Ethnologisches Museum, Staatliche Museen zu Berlin.

62 Sacred groves are common in Ghana. Often these are inhabited by monkeys. It is possible that there are double layers in this fable that I cannot detect, for instance, related to other instances of colonial violence with regard to sacred groves and the monkeys that lived in them, who are usually protected and fed.

AFTERWORD: KNOWING BY EAR

1 The POW recordings of the Lautarchiv are not the first collection of sound recordings I have worked on. See Hoffmann 2009a, 2009b; and Berner, Hoffmann, and Lange 2011.

2 For a brief, recent overview on sound in historiography, see Missfelder 2015. For more on earwitnessing, see Benjamin [1932] 1978; Canetti 1974; Rosenfeld 2011; and Mhlambi 2008, among others. On earwitnessing in museums, see Bijsterfeld 2015.

3 See "War and Grammar: Mohamed Nur: Audio-Visual Traces from the Colonial Archive," MARKK, October 23, 2019–February 23, 2020, https://markk-hamburg.de/en/ausstellungen/war-and-grammar/; and "Der Krieg und die Grammatik/War and Grammar Part I," Anette Hoffman, accessed June 19, 2023, https://anettehoffmann.com/der-krieg-und-die-grammatik-i-2019/.

4 On listening as a practice in the literature of sound studies, see, for instance, Birdsall 2012; and Rice 2015. On close listening, see Hoffmann 2020a, 2021; and Lange 2020. On hearing as different from listening, see Sterne 2015.

5 On my own work on this collection, see "What We See," Anette Hoffmann, accessed February 2022, https://anettehoffmann.com/what-we-see-2009-2013/. See also Hoffmann 2009a, 2009b, 2011a, 2011b.

6 "Decolonizing Southeast Asian Sound Archives (DeCoSEAS)," meLê Yamomo, accessed June 19, 2023, https://site.meleyamomo.com/portfolio/decolonizing-southeast-asian-sound-archives-decoseas/. See also Yamomo 2021, 2018; Titus and Yamomo 2021.

7   The Humboldt University in Berlin is one such example.
8   "Anette Hoffmann," Bergen Assembly 2019, accessed June 19, 2023, https://2019.bergenassembly.no/contributors/anette-hoffmann/.
9   For Nashilongeshipwe Mushaandja, see "Tschukutschuku," Soundcloud, accessed in August 2023, https://soundcloud.com/tschuku-tschuku. For meLê Yamomo, see "Echoing Europe: Postcolonial Reverberations," Ballhaus Naunynstrasse, accessed June 19, 2023, https://ballhausnaunynstrasse.de/play/echoing_europe_-ll__postcolonial_reverberations/. For Memory Biwa and Robert Machiri, see "Untraining the Listening at Pungwe," Savy Contemporary, accessed June 19, 2023, https://savvy-contemporary.com/en/events/2019/untraining-the-ear-listening-session-pungwe/. For Britta Lange and Philip Scheffner, see their website The Halfmoon Files, at https://halfmoonfiles.de/de. For William Kentridge, Philip Miller, Thuthuka Sibisi, and Gregory Maqoma, see "The Head and the Load," Kentridge Studio, 2018, https://www.kentridge.studio/projects/the_head_and_the_load/.

References

Abbink, Jon. 2003. "Somali Dervishes, *Moryaan* and Freedom Fighters: Cycles of Rebellion and the Fragmentation of Somali Society, 1900-2000." In *Rethinking Resistance: Revolt and Violence in African History*, edited by Jon Abbink, Miriam E. de Bruijn, and Klaas van Walraven, 238-66. Leiden: Brill.

Abu-Lughod, Lila, and Catherine Lutz. 1990. "Introduction: Emotion, Discourse and the Politics of Everyday Life." In *Language and the Politics of Emotion*, edited by Lila Abu-Lughod and Catherine Lutz, 1-23. Melbourne: Cambridge University Press.

Afelevo, Stefan H. Kwadzo, and Diedrich Westermann. 1930. "Ein Bericht über den Yehwekultus der Ewe." *Mitteilungen des Seminars für Orientalische Sprachen an der Friedrich-Wilhelms-Universität zu Berlin* 33: 1-55.

Ajotikar, Rasika, and Eva Maria Alexandra van Straaten. 2021. "Postcolonial Sound Archives: Challenges and Potentials; an Introduction." *World of Music* 10 (1): 5-20.

Ames, Eric. 2003. "The Sound of Evolution." *Modernism/Modernity* 10 (2): 297-325.

Andrzejewski, B. W. 2011. "The Literary Culture of Somali People." *Journal of African Cultural Studies* 23 (1): 9-17.

Ankermann, Bernhard. 1914. *Anleitung zum ethnographischen Beobachten und Sammeln*. Berlin: Georg Reimer.

Apoh, Wazi. 2020. "'If Suitable Enforce the Death Penalty': German Solution to Facilitating Colonial Rule in Kete-Krachi." In *Jenseits von Dichotomien/Beyond Dichotomies: Aspekte von Geschichte, Gender, und Kultur in Afrika und Europa/Aspects of History, Gender and Culture in Africa and Europe*, edited by Samuel Ntewusu and Nina Paarmann, 443-66. Berlin: Lit.

Arich-Gerz, Bruno, and Christian Effing. 2017. *Geheimsprachen: Geschichte und Gegenwart verschlüsselter Kommunikation*. Wiesbaden: Marix.

Bakonzi, Agayo. 1982. "The Goldmines of Kilo-Moto in Northeastern Zaire: 1905-1960." PhD diss., University of Wisconsin-Madison.

Bank, Andrew. 2006. *Bushmen in a Victorian World: The Remarkable Story of the Bleek-Lloyd Collection of Bushman Folklore*. Cape Town: Double Storey.

Banti, Giorgio, and Francesco Giannattasio. 1996. "Music and Metre in Somali Poetry." In *Voice and Power: The Culture of Language in North-East Africa; Essays in Honor of B. W. Andrzejewsky*, edited by R. J. Hayward and I. M. Lewis, 83-127. London: School of Oriental and African Studies.

Barnes, Cedric. 2006. "Gubo—Ogadeen Poetry and the Aftermath of the Dervish Wars." *Journal of African Cultural Studies* 18 (1): 105-17.

Barthes, Roland. 1979. *Was singt mir, der ich höre in meinem Körper das Lied*. Translated by Peter Geble. Berlin: Merve.

Baumann, Max Peter. 1997. "Preface: Hearing and Listening in Cultural Context." In "Cultural Concepts of Hearing and Listening." Special issue, *World of Music* 39 (2): 3–8.

Bechhaus-Gerst, Marianne. 2003. *Die koloniale Begegnung: AfrikanerInnen in Deutschland 1880–1945 — Deutsche in Afrika 1880–1918*. Frankfurt: Peter Lang.

Benjamin, Walter. [1932] 1978. *Reflections: Essays, Aphorisms, Biographical Writings*. Translated by Edmund Jephcott. New York: Harcourt Brace Jovanovich.

Berner, Margit. 2003. "Die 'rassenkundlichen' Untersuchungen der Wiener Anthropologen in Kriegsgefangenenlagern 1915–1918." *Zeitgeschichte* 30 (11): 124–36.

Berner, Margit. 2004. "Rassenforschung an kriegsgefangenen Schwarzen." In *Zwischen Charleston und Stechschritt: Schwarze im Nationalsozialismus*, edited by Peter Martin and Christine Alonzo, 605–13. Munich: Dölling und Galitz.

Berner, Margit. 2010. "Large-Scale Anthropological Surveys in Austria-Hungary, 1871–1918." In *Doing Anthropology in Wartime and War Zones: World War I and the Cultural Sciences in Europe*, edited by Reinhard Johler, Christian Marchetti, and Monique Scheer, 233–54. Bielefeld: Transcript. https://pdfs.semanticscholar.org/3a8e/448bb8e61d4652ff1d3d1af47ccf57358fa7.pdf.

Berner, Margit, Anette Hoffmann, and Britta Lange, eds. 2011. *Sensible Sammlungen: Aus dem anthropologischen Depot*. Hamburg: Philo Fine Arts.

Bijsterveld, Karin. 2008. *Mechanical Sound Technology, Culture, and Public Problems of Noise in the Twentieth Century*. Cambridge, MA: MIT Press.

Bijsterveld, Karin. 2015. "Beyond Echoic Memory." *Public Historian* 37 (4): 7–13.

Birdsall, Carolyn. 2012. *Nazi Soundscapes: Sound, Technology and Urban Space in Germany, 1933–1945*. Amsterdam: Amsterdam University Press.

Birdsall, Carolyn, and Viktoria Tkaczyk. 2019. "Listening to the Archive: Sound Data in the Humanities and Sciences." *Technology and Culture* 60 (2S): S1–S13. https://doi.org/10.1353/tech.2019.0061.

Bleek, Wilhelm. 1868. *On the Origins of Language*. New York: L. W. Schmidt.

Boas, Franz. 1911. *Handbook of American Indian Languages, Part I*. Washington, DC: Government Printing Office.

Boas, Franz. 1914. *Kultur und Rasse*. Leipzig: Veit.

Bose, Fritz. 1943/44. "Klangstile als Rassenmerkmale." *Zeitschrift für Rassenkunde* 14: 78–97, 208–24.

Bourdieu, Pierre. 1991. *Language and Symbolic Power*. Translated by Gino Raymond and Matthew Adamson. Cambridge: Polity.

Brady, Erika. 1999. *A Spiral Way: How the Phonograph Changed Ethnography*. Jackson: University Press of Mississippi.

Britton, Celia. 1999. *Edouard Glissant and Postcolonial Theory: Strategies of Language and Resistance*. Charlottesville: University of Virginia Press.

Bronfman, Alejandra. 2016. *Isle of Noise: Media in the Caribbean*. Chapel Hill: University of North Carolina Press.

Brosens, Griet. 2013. *Congo aan den Yser: De 32 Congolese soldaten van het Belgisch leger in de Eerste Wereldoorlog*. Amsterdam: Manteau.

Bull, Michael, and Marcel Cobussen, eds. 2020. *The Bloomsbury Handbook of Sonic Methodologies*. New York: Bloomsbury.

Burkard, Benedikt, and Céline Lebret, eds. 2015. *Gefangene Bilder: Wissenschaft und Propaganda im Ersten Weltkrieg*. Frankfurt: Michael Imhoff.

Canetti, Elias. 1974. *Der Ohrenzeuge: Fünfzig Charaktere*. Munich: Carl Hauser.

Cavarero, Adriana. 2007. "For More Than One Voice." In *Voice and Void*, edited by Thomas Trummer, 44–57. Ridgefield: Aldrich Contemporary Art Museum.

Césaire, Aimé. [1959] 2017. *Über den Kolonialismus*. Translated and with an introduction by Heibert Becker. Berlin: Alexander.

Césaire, Aimé. 1990. *Lyric and Dramatic Poetry 1946–82*. Charlottesville: University of Virginia Press.

Chela, Efemia. 2018. "'Kasala for Myself': New Jazz-Suffused Poetry from Fiston Mwanza Mujila Featured in *Asymptote*, in English and French." *Johannesburg Review of Books*, August 6, 2018, https://johannesburgreviewofbooks.com/2018/08/06/kasala-for-myself-new-jazz-suffused-poetry-from-fiston-mwanza-mujila-featured-in-asymptote-in-english-and-french/.

Chernoff, John. 1997. "Hearing in West African Idioms." In "Cultural Concepts of Hearing and Listening." Special issue, *World of Music* 39 (2): 19–25.

Cohen, Israel. 1917. *The Ruhleben Prison Camp: A Record of Nineteen Months' Internment*. London: Methuen.

Das, Santanu, ed. 2011a. *Race, Empire and First World War Writing*. Cambridge: Cambridge University Press.

Das, Santanu. 2011b. "The Singing Subaltern." *Parallax* 17 (3): 4–18.

Descamps, Cyr, Françoise Descamps, Pierre Rosière, and Guy Thilmans, eds. 2014. *Tirailleurs sénégalais: Témoignages épistolaires, 1914–1919*. Dakar: Éditions des Centaure.

Deumert, Ana, and Nkululeko Mabandla. 2018. "Beyond Colonial Linguistics: The Dialectics of Control and Resistance in the Standardization of isiXhosa." In *Standardizing Minority Languages: Competing Ideologies of Authority and Authenticity in the Global Periphery*, edited by Pia Lane, James Costa, and Haley de Korne, 200–221. New York: Routledge.

Deumert, Ana, Anne Storch, and Nick Shepherd, eds. 2020. *Colonial and Decolonial Linguistics: Knowledges and Epistemes*. Oxford: Oxford University Press.

Deutsches Historisches Museum, ed. 2016. *Deutscher Kolonialismus: Fragmente seiner Geschichte und Gegenwart*. Darmstadt: Theiss.

Diallo, Bakary. [1926] 1985. *Force Bonté*. Paris: Les Nouvelles Éditions Africaines.

Diallo, Bakary, and Lamine Senghor. 2021. *White War, Black Soldiers: Two African Accounts of World War I*. Translated by Nancy Erber and William Peniston. Edited and with an introduction and annotations by George Robb. Indianapolis: Hackett.

Diop, David. 2014. *Frère d'Âme*. Paris: Éditions du Seuil.

Doegen, Wilhelm. 1909. *Doegens Unterrichtshefte für die selbstständige Erlernung fremder Sprachen mit Hilfe der Lautschrift und der Sprechmaschine*. Berlin: O. Schwartz.

Doegen, Wilhelm. 1918. "Denkschrift über die Errichtung eines 'Deutschen Lautamtes' in Berlin." Unpublished manuscript. Staatsbibliothek zu Berlin. https://soundandscience.de/node/912.

Doegen, Wilhelm. 1925. *Unter fremden Völkern: Eine neue Völkerkunde.* Berlin: Otto Stollberg.

Doegen, Wilhelm. 1941. *Unsere Gegener, damals und heute: Engländer und Franzosen mit ihren fremdrassigen Hilfsvölkern in der Heimat, an der Front und in deutscher Gefangenschaft im Weltkriege und im jetzigen Kriege; Großdeutschlands koloniale Sendung.* Berlin: O. F. Huebner.

Dolar, Mladen. 2006. *A Voice and Nothing More.* Cambridge, MA: MIT Press.

Dolar, Mladen. 2007. "Six Lessons in Voice and Meaning." In *Voice and Void*, edited by Thomas Trummer, 28–44. Ridgefield, CT: Aldrich Contemporary Art Museum.

Dreesbach, Anne. 2005. *Gezähmte Wilde: Die Zuschaustellung exotischer Menschen in Deutschland 1870–1940.* Frankfurt: Campus.

Eickstedt, Egon. 1921. "Rassenelement der Sikh." *Zeitschrift für Ethologie* 52: 317–94.

Ellison, Wallace. 1918. *Escaped! Adventures in German Captivity.* Edinburgh: W. Blackwood and Sons.

Erlmann, Veit. 1995. *Nightsong: Performance, Power, and Practice in South Africa.* Chicago: University of Chicago Press.

Erlmann, Veit. 2010. *Reason and Resonance: A History of Modern Aurality.* Cambridge, MA: Zone.

Errington, Joseph. 2008. *Linguistics in a Colonial World: A Story of Language, Meaning, and Power.* Malden, MA: Blackwell.

Eschenberg, Myron. 1991. *Colonial Conscripts: The Tirailleurs Sénégalais in French West Africa, 1857–1960.* London: Portsmouth.

Esselborn, Stefan. 2018. *Die Afrikaexperten: Das Internationale Afrikainstitut und die europäische Afrikanistik, 1926–1976.* Göttingen: Vandenhoek und Ruprecht.

Evans, Andrew D. 2002. "Capturing Race: Anthropology and Photography in German and Austrian Prisoner-of-War Camps during World War I." In *Colonialist Photography: Imag(in)ing Race and Place*, edited by Eleanor M. Hight and Gary D. Sampson, 226–56. London: Routledge.

Evans, Andrew D. 2003. "Anthropology at War: Racial Studies of POWs during World War I." In *Worldly Provincialism: German Anthropology in the Age of Empire*, edited by H. Glenn Penny and Matti Bunzl, 198–229. Ann Arbor: University of Michigan Press.

Evans, Andrew D. 2010a. "Science behind the Lines: The Effect of World War I on Anthropology in Germany." In *Doing Anthropology in Wartime and War Zones: World War I and the Cultural Sciences in Europe*, edited by Reinhard Johler, Christian Marchetti, and Monique Scheer, 99–122. Bielefeld: Transcript.

Evans, Andrew D. 2010b. *Anthropology at War: World War I and the Science of Race in Germany.* Chicago: University of Chicago Press.

Fanon, Frantz. 1965. *A Dying Colonialism.* Translated by Haakon Chevalier. New York: Monthly Review.

Fayola, Toyin. 2016. "Ritual Archives." Lecture presented at the Thabo Mbeki African Leadership Institute, October 27, 2016. YouTube, 1 hr 5 min, accessed June 7, 2023. https://www.youtube.com/watch?v=LAQzPR_5zpg.

Feld, Steven. 1996. "Waterfalls of Song: An Acoustemology of Place in Bosavi, Papua New Guinea." In *Senses of Place*, edited by Steven Feld and Keith H. Basso, 92–137. Santa Fe, NM: School of American Research Press.

Feld, Steven. 2015."Acoustemology." In *Keywords in Sound*, edited by David Novak and Matt Sakakeeny, 12–22. Durham, NC: Duke University Press.

Felderer, Brigitte. 2004. *Phonorama: Eine Kulturgeschichte der Stimme als Medium*. Berlin: Matthes and Seitz.

Finnegan, Ruth. 2012. *Oral Literature in Africa*. Cambridge: Open Book.

Fogarty, Richard S. 2008. *Race and War in France: Colonial Subjects in the French Army, 1914–1918*. Baltimore, MD: Johns Hopkins University Press.

Fogarty, Richard S. 2014. "Out of North Africa: Contested Visions of French Muslim Soldiers during World War I." In *Empires in World War I: Shifting Frontiers and Imperial Dynamics in a Global Conflict*, edited by Andrew Tait Jarboe and Richard S. Fogarty, 136–58. London: I. B. Tauris.

Förster, Larissa. 2010. "Nichts gewagt, nichts gewonnen: Die Ausstellung 'Anders zur Welt kommen. Das Humboldt-Forum im Schloß: Ein Werkstattblick.'" *Paideuma* 56: 241–61.

Foucault, Michel. 1967. "Lives of Infamous Men." In *Power*. Vol. 3, *The Essential Works of Michel Foucault, 1954–1988*, edited by James D. Faubion, translated by Robert Hurley, 157–75. New York: New Press.

Foucault, Michel. [1972] 1996. *Die Ordnung des Diskurses*. Translated by Walter Seitter. Frankfurt: Fischer.

Frobenius, Leo. 1916. *Der Völkerzirkus unserer Feinde*. Foreword by Leo Frobenius. Berlin: Eckart.

Frobenius, Leo, and H. F. von Freytag-Loringhoven. [ca. 1925]. *Deutschlands Gegner im Weltkrieg*. Berlin-Grunewald: Hermann Klemm.

Garcia, Miguel A. 2017. "Sound Archives under Suspicion." In *Historical Sources of Ethnomusicology in Contemporary Debate*, edited by Susanne Ziegler, Ingrid Akesson, Gerda Lechleitner, and Susana Gardo, 10–20. Cambridge: Cambridge Scholars.

Gehrmann, Petra. 2006. "Die Wiederholungsstimme: Über die Strafe der Echo." In *Stimme*, edited by Doris Kolesch and Sybille Krämer, 85–111. Frankfurt: Suhrkamp.

Gerhardt, Ludwig. 2004. "Afrikanistische Forschung: Die Geschichte einer Kontinuität." In *Zwischen Charleston und Stechschritt: Schwarze im Nationalsozialismus*, edited by Christine Alonzo and Peter Martin, 434–50. Munich: Dölling und Galitz.

Gilroy, Paul. 1993. *The Black Atlantic: Modernity and Double Consciousness*. London: Verso.

Gitelman, Lisa. 2004. "Recording Sound, Recording Race, Recording Property." In *Hearing History: A Reader*, edited by Mark M. Smith, 279–94. Athens: University of Georgia Press.

Glissant, Édouard. 1981. *Le discours antillais*. Paris: Éditions du Seuil.

Glissant, Édouard. 2010. *Poetics of Relation*. Translated by Betsy Wing. Ann Arbor: University of Michigan Press.

Grabert, Werner. 1913. "Untersuchungen an Herero- und Hottentotten-Kehlköpfen." *Zeitschrift für Morphologie und Anthropologie* 16 (1): 65–94.

Grierson, George Abraham, comp. and ed. 1903-28. *Linguistic Survey of India*, 11 vols. Calcutta: Office of the Superintendent of Government Printing. https://dsal.uchicago.edu/books/lsi/.

Guha, Ranajit. 1987. "Chandra's Death." In *Subaltern Studies V: Writings on South Asian History and Society*, edited by Ranajit Guha, 135-65. Delhi: Oxford University Press.

Gusinde, Martin. 1948. *Urwaldmenschen am Ituri: Anthropo-biologische Forschungsergebnisse bei Pygmäen und Negern im östlichen Belgisch-Kongo aus den Jahren 1934-35*. Vienna: Springer.

Guthmann, Johannes. 1920. *Scherz und Laune: Max Slevogt und seine Gelegenheitsarbeiten*. Berlin: Paul Cassirer.

Hamilton, Carolyn. 2011. "Backstory, Biography, and the Life of the James Stuart Archive." *History in Africa* 38: 319-41.

Hamilton, Carolyn. 2013. "Forged and Continually Refashioned in the Crucibles of Ongoing Social and Political Life: Archives and Custodial Practices as Subjects of Inquiry." *South African Historical Journal* 65 (1): 1-22.

Hamilton, Carolyn, Verne Harris, Jane Taylor, Michele Pickover, Graeme Reid, and Razia Saleh, eds. 2002. *Refiguring the Archive*. Cape Town: David Philip.

Hamilton, Carolyn, and Nessa Leibhammer. 2014. "Salutes, Labels, and Other Archival Artefacts." In *Uncertain Curature: In and Out of the Archive*, edited by Carolyn Hamilton and Pippa Skotnes, 155-88. Johannesburg: Jacana.

Haraway, Donna. 2016. *Staying with the Trouble: Making Kin in the Chthulucene*. Durham, NC: Duke University Press.

Harrison, Rodney, Sarah Byrne, and Anne Clarke. 2013. *Reassembling the Collection: Ethnographic Museums and Indigenous Agency*. Santa Fe, NM: School of American Research Press.

Heepe, Martin. 1920. *Die Komorendialekte Ngazidja, Nzwani und Mwali*. Hamburg: Hamburger Institut für Kolonialsprachen.

Heinitz, Wilhelm. 1941. "Rassische Merkmale an afrikanischem Musikgut." *Zeitschrift für Rassenkunde und die gesamte Forschung am Menschen* 12: 9-19.

Heller-Roazen, Daniel. 2005. *Echolalien: Über das Vergessen von Sprache*. Translated by Michael Bischoff. Frankfurt: Suhrkamp.

Heller-Roazen, Daniel. 2013. *Dark Tongues: The Art of Rogues and Riddlers*. New York: Zone.

Hermann, E. [1945] 1976. "Schallsignalsprachen in Melanesien und Africa" (1945). Part 1. In *Speech Surrogates: Drum and Whistle Systems*, edited by Thomas A. Sebeok and Donna Jean Umiker-Sebeok, 62-118. The Hague: Mouton.

Hiddleston, J. 2007. "Spivak's Echo: Theorizing Otherness and the Space of Response." *Textual Practice* 21 (4): 623-40.

Hilden, Irene. 2015. "Die (Un)möglichkeit Subalterner Artikulation: Zu den Tonaufnahmen aus deutschen Kriegsgefangenenlagern des Ersten Weltkriegs." MA thesis, Humboldt University, Berlin.

Hilden, Irene. 2021. "Collective Listening: Tracing Colonial Sounds in Postcolonial Berlin." In *The Minor on the Move: Doing Cosmopolitanisms*, edited by Kylie Crane, Lucy Glasser, Sara Morais dos Santos Bruss, and Anna von Rath, 200-223. Münster: Edition Assemblage.

Hilden, Irene. 2022. *Absent Presences in the Colonial Archive: Dealing with the Berlin Sound Archive's Acoustic Legacies.* Leuven: Leuven University Press.

Hilden, Irene, Harriet Merrow, and Andrei Zavadski. 2021. "Present Imperfect, Future Intense: The Digital Opening of the Humboldt Forum." *Boas Blog DCNtR*, March 3, 2021. https://boasblogs.org/dcntr/present-imperfect-future-intense.

Hirschkind, Charles. 2006. *The Ethical Soundscape: Cassette Sermons and Islamic Counterpublics.* New York: Columbia University Press.

Hochschild, Adam. 1999. *King Leopold's Ghost: A Story of Greed, Terror, and Heroism in Colonial Africa.* London: Macmillan.

Hoehne, Markus. 2016. "Dervish State (Somali)." In *The Encyclopedia of Empire*, edited by John M. Mackenzie, 1–2. Chichester: John Wiley and Sons.

Hoffmann, Anette. 2009a. "Widerspenstige Stimmen—Unruly Voices." In *What We See: Reconsidering an Anthropometrical Collection: Images, Voices and Versioning*, edited by Anette Hoffmann, 22–58. Basel: Basler Afrika Bibliographien.

Hoffmann, Anette. 2009b. "Finding Words (of Anger)." In *What We See: Reconsidering an Anthropometrical Collection: Images, Voices and Versioning*, edited by Anette Hoffmann, 114–45. Basel: Basler Afrika Bibliographien.

Hoffmann, Anette. 2011a. "Glaubwürdige Inszenierungen: Die Produktion von Abformungen in der Polizeistation von Keetmanshoop im August 1931." In *Sensible Sammlungen: Aus dem anthropologischen Depot*, edited by Margit Berner, Anette Hoffmann, and Britta Lange, 61–88. Hamburg: Philo Fine Arts.

Hoffmann, Anette. 2011b. "Voice Over II: 'Wie ein Hund in einem Fangeisen schreien.'" In *Sensible Sammlungen: Aus dem anthropologischen Depot*, edited by Margit Berner, Anette Hoffmann, and Britta Lange, 129–46. Hamburg: Philo Fine Arts.

Hoffmann, Anette. 2012. "Chronotopes of the (Post-)Colonial Condition in Otjiherero Praise Poetry." In *Style in African Literature*, edited by J. K. S. Makokha, Ogone John Obiero, and Russel West-Pavlov, 309–27. Leiden: Brill.

Hoffmann, Anette. 2014a. "Echoes of the Great War: The Recordings of African Prisoners in the First World War." *Open Arts Journal* 3: 9–23.

Hoffmann, Anette. 2014b. "Of Storying and Storing: 'Reading' Lichtenecker's Voice Recordings." In *Re-Viewing Resistance in Namibian History*, edited by Jeremy Silvester, 89–104. Windhoek: UNAM Press.

Hoffmann, Anette. 2015a. "Listening to Sound Archives: Introduction to Edited Section." *Social Dynamics* 41 (1): 73–83.

Hoffmann, Anette. 2015b. "Verbal Riposte: Wilfred Tjiueza's Performances of Omitandu as Responses to the Racial Model of Hans Lichtenecker." In *Acts of Voicing*, edited by Hans D. Christ, Iris Dressler, and Christine Peters, 323–32. Leipzig: Spector.

Hoffmann, Anette. 2018. "Kolonialität." In *Handbuch Sound: Geschichte-Begriffe-Ansätze*, edited by Daniel Morat and Hansjakob Ziemer, 387–90. Stuttgart: Metzler.

Hoffmann, Anette. 2019. "'Achtung Aufnahme!' Akustische Spuren der kolonialen Wissensproduktion" In *Museumsethnologie—Eine Einführung: Theorien-Debatten-Ansätze*, edited by Iris Edenheiser and Larissa Förster, 204–6. Berlin: Dietrich Reimer.

Hoffmann, Anette. 2020a. *Kolonialgeschichte hören: Das Echo gewaltsamer Wissensproduktion in Tondokumenten aus dem südlichen Afrika*. Vienna: Mandelbaum.

Hoffmann, Anette. 2020b. "War and Grammar: Acoustic Recordings with African Prisoners of the First World War (1915-18)." In *Colonial and Decolonial Linguistics: Knowledges and Epistemes*, edited by Ana Deumert, Nick Shepherd, and Anne Storch, 105-28. Oxford: Oxford University Press.

Hoffmann, Anette. 2021a. "Close Listening: Approaches to Research on Colonial Sound Archives." In *The Bloomsbury Handbook of Sonic Methodologies*, edited by Marcel Cobussen and Michael Bull, 529-39. New York: Bloomsbury.

Hoffmann, Anette. 2021b. "Skandalträchtig drauflosreden: Vorschlag zur Entsachlichung des Sprechens von der Erbeutung von Körpern, Objekten, und von Praktiken der kolonialen Linguistik in vier Stücken." *The Mouth: Critical Studies on Language, Culture and Society* 9: 11-30.

Hoffmann, Anette. 2023. *Listening to Colonial History: Echoes of Coercive Knowledge Production in Historical Sound Archives from Southern Africa*. Basel: Basler Afrika Bibliographien.

Hoffmann, Anette, and Phindezwa Mnyaka. 2015. "Hearing Voices in the Archive." *Social Dynamics* 41 (1): 101-23.

Honour, Hugh. 2012. "The New Negro." In *The Image of the Black in Western Art*. Vol. 4, *From the American Revolution to World War I, Part 2: Black Models, White Myth*, edited by David Bindman and Henry Louis Gates Jr., 191-246. Cambridge, MA: Harvard University Press.

hooks, bell. 2015. *Yearning: Race, Gender, and Cultural Politics*. New York: Routledge.

Höpp, Gerhard. 1997. *Muslime in der Mark: Als Kriegsgefangene und Internierte in Wünsdorf und Zossen, 1914-1924*. Berlin: Das Arabische Buch.

Hüsgen, Jan. 2020. "Colonial Expeditions and Collecting: The Context of the 'Togo Hinterland Expedition' of 1894/1895." *Journal for Art Market Studies* 1: 1-12.

Imiela, Hans-Jürgen. 1968. *Max Slevogt: Eine Monographie*. Karlsruhe: G. Braun.

Irvine, Judith. 1990. "Registering Affect: Heteroglossia in the Linguistic Expression of Emotion." In *Language and the Politics of Emotion*, edited by Catherina A. Lutz and Lila Abu-Lughod, 126-61. Cambridge: Cambridge University Press; Paris: Éditions de la maison des sciences de l'homme.

Irvine, Judith. 2008. "Subjected Words: African Linguistics and the Colonial Encounter." *Language and Communication* 28: 323-43.

Johler, Reinhard, Christian Marchetti, and Monique Scheer, eds. 2010. *Doing Anthropology in Wartime and War Zones: World War I and the Cultural Sciences in Europe*. Bielefeld: Transcript.

Kahleyss, Margot. 2015. "Kolonialsoldaten in Gefangenschaft und Lager." In *Gefangene Bilder: Wissenschaft und Propaganda im Ersten Weltkrieg*, edited by Benedikt Burkard and Céline Lebret, 24-52. Frankfurt: Michael Imhoff.

Kalibani, Méhéza. 2021. "Kolonialer Tinnitus: Das belastende Geräusch des Kolonialismus." *Geschichte in Wissenschaft und Unterrricht* 9-10: 540-53.

Kaplan, Judith. 2013. "Voices of the People: Linguistic Research among Germany's Prisoners of War during World War I." *Journal for the History of Science* 49 (3): 281-305.

Katjavivi, Peter. 1988. *A History of Resistance in Namibia*. Paris: UNESCO Press.
Ketchum, John D. 1965. *Ruhleben: A Prison Camp Society*. Toronto: University of Toronto Press.
Khumalo, Fred. 2017. *Dancing the Death Drill*. Johannesburg: Penguin Random House South Africa.
Koelle, Sigismund. 1854. *Polyglotta Africana, or, A Comparative Vocabulary of Nearly Three Hundred Words and Phrases in More Than One Hundred Distinct African Languages*. London: Church Missionary House.
Kolesch, Doris, and Sybille Krämer. 2006. "Stimmen im Konzert der Disziplinen: Zur Einführung in diesen Band." In *Stimme: Annäherung und ein Phänomen*, edited by Doris Kolesch and Sybille Krämer, 7–16. Frankfurt: Suhrkamp.
Koller, Christian. 2004. "'Der dunkle Verrat an Europa': Afrikanische Soldaten im Krieg 1914–1918 in der deutschen Wahrnehmung." In *Zwischen Charleston und Stechschritt: Schwarze im Nationalsozialismus*, edited by Christine Alonzo and Peter Martin, 111–15. Munich: Dölling und Gallitz.
Koller, Christian. 2008. "The Recruitment of Colonial Troops in Africa and Asia and Their Deployment in Europe during the First World War." *Immigrants and Minorities* 26 (1–2): 111–33.
Koller, Christian. 2011a. "German Perceptions of Enemy Colonial Troops, 1914–1918." In *"When the War Began We Heard of Several Kings": South Asian Prisoners in World War I in Germany*, edited by Franziska Roy, Heike Liebau, and Ravi Ahuja, 130–48. New Delhi: Social Science.
Koller, Christian. 2011b. "Representing Otherness: African, Indian and European Soldiers' Letters and Memoirs." In *Race, Empire and First World War Writing*, edited by Santanu Das, 127–42. Cambridge: Cambridge University Press.
Koller, Christian. 2014. "Colonial Military Participation in Europe (Africa)." In *1914-1918 Online: International Encyclopedia of the First World War*. https://encyclopedia.1914-1918-online.net/article/colonial_military_participation_in_europe_africa?version=1.0.
Kongolo, Antoine Tshitungu. 2011. *Visages de Paul Panda Farnana: Nationaliste, panafricaniste, intellectuel engagé*. Paris: L'Harmattan.
Kracauer, Siegfried. 1995. *The Mass Ornament: Weimar Essays*. Translated, edited, and with an introduction by Thomas Y. Levin. Cambridge, MA: Harvard University Press.
Krzywick, Janusz. 1984. "Contes didaticques Bira (Haut-Zaire)." *Africa: Rivista trimestrale di studi e documentazion dell' Instituto per Africa e l'Oriente* 39 (3): 416–43.
Krzywick, Janusz. 2014. *Contes Bira*. Vol. 1, *Solenyama*. Warsaw: Département des Langues et des Cultures Africaines.
Kuba, Richard. 2015. "Leo Frobenius und der Erste Weltkrieg." In *Gefangene Bilder: Wissenschaft und Propaganda im Ersten Weltkrieg*, edited by Benedikt Burkhardt and Céline Lebret, 102–16. Frankfurt: Michael Imhoff.
Lalu, Premesh. 2009. *The Deaths of Hintsa: Postapartheid South Africa and the Shape of Recurring Pasts*. Cape Town: HSRC Press.
Landau, Paul. 2010. *Popular Politics in the History of South Africa, 1400–1948*. Cambridge: Cambridge University Press.

Lange, Britta. 2010. "Aftermath: Anthropological Data from Prisoner-of-War Camps." In *Doing Anthropology in Wartime and War Zones: World War I and the Cultural Sciences in Europe*, edited by Reinhard Johler, Christian Marchetti, and Monique Scheer, 311–36. Bielefeld: Transcript.

Lange, Britta. 2011a. "South Asian Soldiers and German Academics: Anthropological, Linguistic, and Musicological Studies in Prison Camps." In *"When the War Began We Heard of Several Kings": South Asian Prisoners in World War I Germany*, edited by Franziska Roy, Heike Liebau, and Ravi Ahuja, 149–86. New Delhi: Social Science.

Lange, Britta. 2011b. "'Denken Sie selber über diese Sache nach . . .': Tonaufnahmen in deutschen Gefangenenlagern des Ersten Weltkriegs." In *Sensible Sammlungen: Aus dem anthropologischen Depot*, edited by Margit Berner, Anette Hoffmann, and Britta Lange, 89–129. Hamburg: Fundus.

Lange, Britta. 2013. *Die Wiener Forschungen an Kriegsgefangenen 1915–1918: Anthropologische und ethnographische Verfahren im Lager*. Vienna: OAW Verlag.

Lange, Britta. 2015. "Die Gefangenen als Untersuchungsobjekte." In *Gefangene Bilder: Wissenschaft und Propaganda im Ersten Weltkrieg*, edited by Benedikt Burkard and Céline Lebret, 84–94. Frankfurt: Historisches Museum Frankfurt.

Lange, Britta. 2020. *Gefangene Stimmen: Tonaufnahmen von Kriegsgefangenen aus dem Lautarchiv 1915–1918*. Berlin: Kadmos.

Lange, Britta. 2022. *Captured Voices: Sound Recordings of Prisoners of War from the Sound Archive 1915–1918*. Translated by Rubaica Jakiwala. Berlin: Kadmos.

Lautbibliothek, ed. 1928. *Afrikanische Sprachen: Arabische und Berberische Sprachen*. Phonetische Platten und Umschriften, no. 45. Produced by Hans Stumme. Berlin: Preußische Staatsbibliothek.

Lautbibliothek, ed. 1929a. *Afrikanische Sprachen: Suaheli*. Phonetische Platten und Umschriften, no. 25. Produced by A. Schroeder. Berlin: Preußische Staatsbibliothek.

Lautbibliothek, ed. 1929b. *Mandara*. Phonetische Platten und Umschriften, no. 48. Produced by Carl Meinhof and August Klingenheben. Berlin: Preußische Staatsbibliothek.

Lautbibliothek, and Diedrich Westermann, eds. 1936. *Afrikanische Sprachen: Xhosa*. Phonetische Platten und Umschriften, no. 42. Recorded by Carl Meinhof. Produced by A. Tucker. Berlin: Preußische Staatsbibliothek.

Le Guin, Ursula K. [1988] 2019. *The Carrier Bag Theory of Fiction*. Introduction by Donna Haraway. London: Ignota.

Liebau, Heike. 2011. "The German Foreign Office, Indian Emigrants and Propaganda Efforts among the 'Sepoys.'" In *"When the War Began We Heard of Several Kings": South Asian Prisoners in World War I in Germany*, edited by Franziska Roy, Heike Liebau, and Ravi Ahuja, 96–129. New Delhi: Social Science.

List of Merchant Seamen and Fishermen Detained as Prisoners of War in Germany, Austria-Hungary and Turkey. London: Board of Trade, 1918.

Loez, André. 2017. "Between Acceptance and Refusal: Solders' Attitudes towards War (Africa)." In *1914–1918 Online: International Encyclopedia of the First World War*. Accessed November 2020. https://encyclopedia.1914-1918-online.net/article/between_acceptance_and_refusal_-_soldiers_attitudes_towards_war.

Lombard, Rosemary. 2018. "Das bloße Leben: Koloniale und neokoloniale Abbildungen südafrikanischer Bergarbeiter in der öffentlichen Wahrnehmung." In *Zum Beispiel BASF: Über Konzernmacht und Menschenrechte*, edited by Britta Becker, Maren Grimm, and Jakob Krameritsch, 370–85. Vienna: Mandelbaum.

Lowe, Lisa. 2015. *The Intimacies of Four Continents*. Durham, NC: Duke University Press.

Lunn, Joe. 1987. "Kande Kamara Speaks: An Oral History of the West African Experience in France 1914–18." In *Africa and the First World War*, edited by Melvin Page, 28–53. New York: St. Martin's.

Lunn, Joe. 1999. *Memoirs of the Maelstrom: A Senegalese Oral History of the First World War*. London: James Currey.

Lunn, Joe. 2011. "France's Legacy to Demba Mboup? A Senegalese Griot and His Descendants Remember His Military Service during the First World War." In *Race, Empire and First World War Writing*, edited by Santanu Das, 108–24. Cambridge: Cambridge University Press.

Macchiarelli, Ignazio, and Emilio Tamburini. 2018. *Le voci ritrovate: Canti e narrazioni di prigionieri italiani della Grande Guerra negli archivi sonori di Berlino*. Udine: Nota.

Maier, Donna. 1980. "Competition for Power and Profits in Kete-Krachi, West Africa 1875–1900." *International Journal of African Historical Studies* 19 (1): 33–35.

Mangin, Charles. 1910. *La force noire*. Paris: Librarie Hachette et Cie.

Marchal, Jules. 2003. *Forced Labor in the Gold and Copper Mines: A History of Congo under Belgian Rule, 1910–1945*, translated from the French by Ayi Kwei Armah. Popenguine, Senegal: Per Ankh.

Matiasek, Katarina. 2017. "Rudolf Poech's 'Revenants': Photography and Atavism in Austro-Hungarian Prisoner-of-War Studies 1915–1918." *International Forum on Audio-Visual Research: Jahrbuch des Phonogrammarchivs* 7: 30–45. https://doi.org/10.1553/jpa7s30.

Mbembe, Achille. 2014. *Kritik der schwarzen Vernunft*. Translated by Michael Bischoff. Frankfurt: Suhrkamp.

Meinhof, Carl. [1894] 1976. "Die Geheimsprachen Afrikas." In *Speech Surrogates: Drum and Whistle Systems*, edited by Thomas A Sebeok and Donna Jean Umiker Sebeok, 1:151–56. The Hague: Mouton.

Meinhof, Carl. 1905. *Die Christianisierung der Sprachen Afrikas*. Basel: Basler Missionsstudien.

Meinhof, Carl. 1911. *Die Dichtung der Afrikaner: Hamburgische Vorträge*. Berlin: Buchhandlung der Berliner Missionsgesellschaft.

Meinhof, Carl. 1912. *Die Sprache der Hamiten*. Abhandlungen des Hamburgischen Kolonialinstitutes 9. Hamburg: Friedrichson.

Meinhof, Carl. 1913. "Das Evangelium und die primitive Rassen." In *Biblische Zeit—Und Streitfragen zur Aufklärung der Gebildeten*, edited by Friedrich Kropatschek. Berlin: Erwin Runge.

Meinhof, Carl. 1938. "Die Enstehung der Bantusprachen." *Zeitschrift für Ethnologie* 70 (3-4): 144–52.

Meinhof, Carl. 1939. "Die Sprache der Bira." *Zeitschrift für Eingeborenensprachen* 29: 241–87.

Mergenthaler, Volker. 2007. *Völkerschau-Kannibalismus-Fremdenlegion*. Tübingen: Max Niemeyer.

Meyer, Birgit. 2002. "Christianity and the Ewe Nation: German Pietist Missionaries, Ewe Converts and the Politics of Culture." *Journal of Religion in Africa* 32: 167–99.

Meyer-Bahlburg, Hilke, and Ekkehard Wolff. 1986. *Afrikanische Sprachen in Forschung und Lehre: 75 Jahre Afrikanistik in Hamburg (1909–1984)*. Hamburg: Reimer.

Mhlambi, Thokozani Ndumiso. 2008. "The Early Years of Black Radio Broadcasting in South Africa: A Critical Reflection on the Making of Ukhosi FM." PhD diss., University of Cape Town.

Mignolo, Walter. 2007. "Coloniality and Modernity/Rationality." *Cultural Studies* 21 (2): 155–67.

Missfelder, Jan-Friedrich. 2012. "Period Ear: Perspektiven einer Klanggeschichte der Neuzeit." *Geschichte der Gegenwart* 38: 21–47.

Missfelder, Jan-Friedrich. 2015. "Geschichtswissenschaft." In *Handbuch Sound: Geschichte-Begriffe-Ansätze*, edited by Daniel Morat and Hansjakob Ziemer, 107–12. Stuttgart: Metzler.

Morat, Daniel, ed. 2014. *Sounds of Modern History: Auditory Cultures in 19th- and 20th-Century Europe*. New York: Berghahn.

Morrison, Toni. 1993. *Playing in the Dark: Whiteness and the Literary Imagination*. New York: Vintage.

Morrison, Toni. 2019. *Mouth Full of Blood: Essays, Speeches, Meditations*. London: Penguin.

Mowitt, John. 1992. *Text: The Genealogy of an Antidisciplinary Object*. Durham, NC: Duke University Press.

Murrell, Denise. 2018. *Posing Modernity: The Black Model from Manet and Matisse to Today*. Exhibition catalogue. New Haven, CT: Yale University Press in association with the Miriam and Ira D. Wallach Art Gallery, Columbia University.

Nancy, Jean-Luc. 2007. *Listening*. New York: Fordham University Press.

Njung, George N. 2014. "#GreatWarinAfrica: Honour Motivated Some Cameroonian Soldiers Who Fought for Germany in the First World War." *Africa at LSE* (blog), August 26, 2014. https://blogs.lse.ac.uk/africaatlse/2014/08/26/greatwarinafrica-honour-was-a-motivating-factor-for-some-cameroonian-soldiers-who-fought-for-germany-during-the-first-world-war/.

Njung, George N. 2020. "Amputated Men, Colonial Bureaucracy, and Masculinity in Post-World War I Colonial Nigeria." *Journal of Social History* 53 (3): 620–43.

Nketia, J. H. Kwabena. 1963. *Drumming in Akan Communities of Ghana*. Legon: University of Ghana Press.

Ntewusu, Samuel Aniegeye. 2015. "The Impact and Legacies of German Colonialism in Kete Krachi, North-Eastern Ghana." ASC Working Paper 121. Leiden: African Studies Centre.

Ochoa Gautier, Ana María. 2014. *Aurality: Listening and Knowledge in Nineteenth-Century Columbia*. Durham, NC: Duke University Press.

Oloukpona-Yinnon, Adjaï Paulin. 1998. *Unter deutschen Palmen: Die "Musterkolonie Togo" im Spiegel deutscher Kolonialliteratur*. Frankfurt: IKO Verlag.

Orwin, Martin. 2005. "On the Concept of 'Definitive Text' in Somali Poetry." *Oral Tradition* 20(2): 278–99.

Page, Melvin E. 1987. "Introduction: Black Men in a White Man's War." In *Africa and the First World War*, edited by Melvin E. Page, 1–27. New York: St. Martin's.

Pandey, Gyanendra. [2000] 2012. "Voices from the Edge: The Struggle to Write Subaltern Histories." In *Mapping Subaltern Studies and Postcolonialism*, edited by Vinayak Chaturvedi, 281–99. London: Verso.

Peeren, Esther. 2009. "Seeing More (Hi)stories: Versioning as Resignificatory Practice in the What We See Exhibition and the Work of Sanell Aggenbach and Mustafa Maluka." In *What We See: Reconsidering an Anthropometrical Collection from Southern Africa*, edited by Anette Hoffmann, 84-104. Basel: Basler Afrika Bibliographien.

Pfäffinger, Jonas. 2015. "Völkerschauen im deutschen Kaiserreich und die zeitgenössiche Kritik." BA thesis, Philipps Universität, Marburg.

Phadi, Mosa, and Nomancotsho Pakade. 2016. "The Native Informant Speaks Back to the Offer of Friendship in White Academia." In *Ties That Bind: Race and the Politics of Friendship in South Africa*, edited by Shannon Walsh and Jon Soske, 288-307. Johannesburg: Wits University Press.

Pöch, Rudolf. 1916. "Anthropologische Studien an Kriegsgefangenen." *Die Umschau*, no. 20: 988–91.

Prakash, Gyan. 1994. "Subaltern Studies as Postcolonial Criticism." *American Historical Review* 99 (5): 1475–90.

Pugach, Sarah. 2000. "Christianize and Conquer: Carl Meinhof, German Evangelical Missionaries, and the Debate over African Languages 1905-1912." In *Mission und Gewalt: Der Umgang der Christliche Mission mit Gewalt und die Ausbreitung des Christentums in Afrika und Asien in der Zeit von 1792-1918*, edited by Ulrich van der Heyden and Jürgen Becher, 509–24. Stuttgart: Franz Steiner.

Pugach, Sarah. 2012. *Africa in Translation*. Ann Arbor: University of Michigan Press.

Pugach, Sarah. 2018. "A Short History of African Language Studies in the Nineteenth and Early Twentieth Century, with Emphasis on German Contributors." In *The Routledge Handbook of African Linguistics*, edited by Augustine Agwuele and Adams Bodomo, 15-32. London: Routledge.

Quijano, Anibal. 2000. "The Coloniality of Power, Eurocentrism and Latin America." *Nepantla* 1 (3): 533–80.

Quijano, Anibal. 2007. "Coloniality and Modernity/Rationality." *Cultural Studies* 21 (2–3): 168–78.

Quinn, Frederik. 1987. "The Impact of the First World War and Its Aftermath on the Beti of Cameroun." In *Africa and the First World War*, edited by Melvin Page, 171–85. New York: St. Martin's.

Quintero, Pablo, and Sebastian Garbe, eds. 2013. *Kolonialität der Macht*. Hamburg: Unrast.

Radano, Ronald, and Philip von Bohlmann, eds. 2000. "Introduction: Music and Race, Their Past and Their Presence." In *Music and the Racial Imagination*, edited by Ronald Radano and Philip von Bohlmann, 1–56. Chicago: University of Chicago Press.

Rassool, Ciraj. 2006. "Beyond the Cult of 'Salvation' and 'Remarkable Equality': A New Paradigm for the Bleek and Lloyd Collection." *Kronos* 32: 244–51.

Rassool, Ciraj, and Martin Legassick. 2000. *Skeletons in the Cupboard: South African Museums and the Trade in Human Remains 1907-1917*. Cape Town: The South African Museum.

Renton, David, David Seddon, and Leo Zeilig. 2007. *The Congo: Plunder and Resistance*. London: Zed.

Rice, Tom. 2015. "Listening." In *Keywords in Sound*, edited by David Novak and Matt Sakakeeny, 100–108. Durham, NC: Duke University Press.

Riesz, János. 2011. "Afrikanische Kriegsgefangene in deutschen Lagern während des Ersten Weltkriegs." In *Deutsch-Afrikanische Diskurse in Geschichte und Gegenwart: Literatur und kulturwissenschaftliche Perspektiven*, edited by Michael Hofmann and Rita Morrien, 71–106. Amsterdam: Rodopi.

Riva, Nepomuk. 2015. "Ethnologie." In *Handbuch Sound: Geschichte-Begriffe-Ansätze*, edited by Daniel Morat and Hansjakob Ziemer, 102–6. Stuttgart: Metzler.

Robinson, Dylan. 2020. *Hungry Listening: A Resonating Theory for Indigenous Sound Studies*. Minneapolis: University of Minnesota Press.

Römer, Ruth. 1989. *Sprachwissenschaft und Rassenideologie in Deutschland*. Munich: Wilhelm Fink.

Rosenfeld, Sophia. 2011. "On Being Heard: A Case for Paying Attention to the Historical Ear." *American Historical Review* 116 (2): 316–34.

Roy, Franziska, Heike Liebau, and Ravi Ahuja, eds. 2011. *"When the War Began We Heard of Several Kings": South Asian Prisoners in World War I Germany*. New Delhi: Social Science.

Sacken, Katharina. 2004. "Ungern vor Fremden gesungen: Koloniale Phonographie um 1900." In *Phonorama: Eine Kulturgeschichte der Stimme als Medium*, edited by Brigitte Felderer, 118–31. Berlin: Matthes und Seitz.

Samatar, Said S. 1982. *Oral Poetry and Somali Nationalism: The Case of Sayyid Mahammad 'Abdille Hassan*. Cambridge: Cambridge University Press.

Sarreiter, Regina. 2012. "Ich glaube, dass die Hälfte ihres Museums gestohlen ist." In *Was Wir Sehen: Bilder, Stimmen, Rauschen: Zur Kritik anthropometrischen Sammelns*, edited by Anette Hoffmann, Britta Lange, and Regina Sarreiter, 43–60. Basel: Basler Afrika Bibliographien.

Saussure, Ferdinand de. 1916. *Course in General Linguistics*. Translated by Wade Baskin. New York: Philosophical Society.

Schaeffer, Pierre. 2004. "Acousmatics." In *Audio Culture: Readings in Modern Music*, edited by Christopher Cox and Daniel Warner, 76–81. New York: Continuum International.

Schasiepen, Sophie. 2019. "Die 'Lehrmittelsammlung' von Dr. Rudolf Pöch an der Universität Wien: Anthropologie, Forensik und Provenienz." *Zeitschrift für Kulturwissenschaften* 6 (1): 115–28. https://doi.org/10.25969/mediarep/13898.

Schasiepen, Sophie. 2021. "Physical Anthropology and the Production of Racial Capital in Austria." PhD diss., University of the Western Cape, South Africa.

Schebesta, Paul. 1934. *Vollblutneger und Halbzwerge: Forschungen unter Waldnegern und Halbpygmäen am Ituri in Belgisch Kongo.* Salzburg: Anton Pustet.

Schebesta, Paul. 1946/49. "Die Waldneger: Paläonegride und Negro-Bantuide am Ituri (Belg. Kongo)." *Anthropos* 41–44 (1–3): 161–76.

Scheer, Monique. 2010."Captive Voices: Phonographic Recordings in the German and Austrian Prisoner-of-War Camps of World War I." In *Doing Anthropology in Wartime and War Zones: World War I and the Cultural Sciences in Europe,* edited by Reinhard Johler, Christian Marchetti, and Monique Scheer, 279–309. Bielefeld: Transcript.

Schmidt, Leigh Eric. 2000. *Hearing Things: Religion, Illusion, and the American Enlightenment.* Cambridge, MA: Harvard University Press.

Schmidt, Sigrid. 2001. *Tricksters, Monsters and Clever Girls: African Folktales, Texts and Discussions.* Cologne: Rüdiger Köppe.

Schrödl, Jenny, and Doris Kolesch. 2018. "Stimme." In *Handbuch Sound: Geschichte-Begriffe-Ansätze,* edited by Daniel Morat and Hansjakob Ziemer, 223–29. Stuttgart: Metzler.

Scott, David. 2018. "Evil beyond Repair." *Small Axe* 22 (1): vii–x.

Sebald, Peter. 1987. *Togo 1884–1914.* Berlin: Akademie.

Sebeok, Thomas A., and Donna Jean Umiker Sebeok, eds. 1976. *Speech Surrogates: Drum and Whistle Systems.* 2 vols. The Hague: Mouton.

Senghor, Lamine. 2012. *La violation d'un pays et autre ecrits anticolonialistes.* Paris: L'Harmattan.

Senghor, Léopold Sédar. 1984. *Poèmes: Nouvelle édition.* Paris: Éditions du Seuil.

Shaw, George Bernard. [1916] 2003. *Pygmalion: A Romance in Five Acts.* London: Penguin.

Shepherd, Nick. 2015. *The Mirror in the Ground: Archeology, Photography and the Making of a Disciplinary Archive.* Jeppestown: Jonathan Ball.

Simon, Artur, ed. 2000. *Das Berliner Phonogramm-Archiv: Sammlungen der traditionellen Musik der Welt.* Berlin: VBB Verlag.

Smith, Mark M. 2001. *Listening to Nineteenth-Century America.* Chapel Hill: University of North Carolina Press.

Spieth, Jakob. 1911. *Die Religion der Eweer in Süd-Togo.* Göttingen: Vadenhoeck and Ruprecht.

Spivak, Gayatri Chakravorty. 1988. "Can the Subaltern Speak?" In *Marxism and the Interpretation of Cultures,* edited by Laurence Grossberg and George Nelson, 271–313. London: Macmillan.

Spivak, Gayatri Chakravorty. 1993. "Echo." *New Literary History* 25 (1): 17–43.

Spivak, Gayatri Chakravorty. 1999. *A Critique of Postcolonial Reason: Toward a History of the Vanishing Present.* Cambridge, MA: Harvard University Press.

Spivak, Gayatri Chakravorty. 2010. "Translating in a World of Languages." *Profession*: 35–43.

Sporcq, Jacques. 1975. "The Bira of the Savanna and the Bira of the Rain Forest: A Comparative Study of Two Populations of the Democratic Republic of the Congo." *Journal of Human Evolution* 4 (6): 505–16.

Steedman, Carolyn. 2000. "Enforced Narratives: Stories of Another Self." In *Feminism and Autobiography: Texts, Theories, Methods,* edited by Tess Coslet, Celia Lury, and Penny Summerfield, 25–39. London: Routledge.

Sterne, Jonathan. 2003. *The Audible Past: Cultural Origins of Sound Reproduction*. Durham, NC: Duke University Press.

Sterne, Jonathan. 2004. "Preserving Sound in Modern America." In *Hearing History: A Reader*, edited by Mark M. Smith, 295–318. Athens: University of Georgia Press.

Sterne, Jonathan. 2015. "Hearing." In *Keywords in Sound*, edited by David Novak and Matt Sakakeeny, 65–77. Durham, NC: Duke University Press.

Stibbe, Edward V. [1919] 1969. *Reminiscences of a Civilian Prisoner in Germany, 1914–1918*. With a foreword by Paul Stibbe. Castle Cary, UK: N.p., 1969.

Stibbe, Matthew. 2008. *British Civilian Internees in Germany: The Ruhleben Camp, 1914–18*. Manchester: Manchester University Press.

Stiehl, Otto. 1916. *Unsere Feinde: 96 Charakterköpfe aus deutschen Kriegsgefangenenlagern*. Stuttgart: J. Hoffmann.

Stoecker, Holger. 2008. *Afrikawissenschaften in Berlin von 1919 bis 1945: Zur Geschichte und Topographie eines wissenschaftlichen Netzwerkes*. Stuttgart: Franz Steiner.

Stoecker, Holger. 2013. "Lehrer, Informanten, Studienobjekte. Afrikanische Sprachlektoren im Berlin der Zwischenkriegsjahre." In *Black Berlin: Die Deutsche Metropole und Ihre Afrikanischen Diaspora in Geschichte und Gegenwart*, edited by Oumar Diallo and Joachim Zeller, 71–88. Berlin: Metropol.

Stoecker, Holger, Jürgen Mahrenholz, and the Black Europe Team. 2013. "Die Lautabteilung an der Preußischen Staatsbibliothek." In *Black Europe*, Vol. 2, *The First Comprehensive Documentation of the Sounds and Images of Black People in Europe, Pre-1927* (book accompanying CDs), edited by J. Green, R. E. Lotz, and H. Rye, 205–12. Holste-Oldendorf, Germany: Bear Family Productions.

Stoever, Jennifer Lynn. 2016. *The Sonic Color Line: Race and the Cultural Politics of Listening*. New York: New York University Press.

Stoler, Ann Laura. 2008. "Imperial Debris: Reflections on Ruins and Ruination." *Cultural Anthropology* 23 (2): 191–219.

Stoler, Ann Laura. 2009. *Along the Archival Grain: Epistemic Anxieties and Colonial Commonsense*. Princeton, NJ: Princeton University Press.

Stoler, Anne Laura. 2016. *Duress: Imperial Durabilities in Our Times*. Durham, NC: Duke University Press.

Stolz, Thomas, Ingo H. Warnke, and Daniel Schmidt-Brücken, eds. 2016. *Sprache und Kolonialismus: Eine interdisziplinäre Einführung zu Sprache und Kommunikation in kolonialen Kontexten*. Berlin: De Gruyter.

Stopa, Roman. 1935. *Die Schnalze: Ihre Natur, Entwicklung und Ursprung*. Krakow: Polska Akademja Umiejetnosci.

Storch, Anne. 2011. *Secret Manipulations: Language and Context in Africa*. Oxford: Oxford University Press.

Storch, Anne. 2018. "Gewissheit und Geheimnis." In *Sprache und (Post)-Kolonialismus: Linguistische und Interdisziplinäre Aspekte*, edited by Birte Kellermeier-Rehbein, Matthias Schulz, and Doris Stolberg, 105–26. Berlin: De Gruyter.

Struck, Hermann. 1917. *Kriegsgefangene: Hundert Steinzeichnungen von Hermann Struck; Mit Begleitworten von F. von Luschan; Ein Beitrag zur Völkerkunde im Weltkriege.* Berlin: Ernst Vohsen.

Stumpf, Carl. 1911. *Die Anfänge der Musik.* Leipzig: J. A. Barth.

Taylor, Diana. 2003. *The Archive and the Repertoire: Performing Cultural Memory in the Americas.* Durham, NC: Duke University Press.

Thilenius, G., C. Meinhof, and W. Heinitz. [1916] 1976. "Die Trommelsprache in Afrika und in der Südsee." In *Speech Surrogates: Drum and Whistle Systems,* edited by Thomas A. Sebeok and Jean Umiker-Sebeok, 3–32. The Hague: Mouton.

Thode-Aurora, Hilke. 1989. *Für fünfzig Pfennig um die Welt: Die Hagenbeckschen Völkerschauen.* Frankfurt: Campus.

Tracey, Hugh. 1948. *Lalela Zulu: 100 Zulu Lyrics.* Illustrated by Eric Byrd. Johannesburg: African Music Society.

Trouillot, Michel-Rolph. 1995. *Silencing the Past.* Boston: Beacon.

Ulrikab, Abraham, Alootook Ilellie, Hans-Ludwig Blohm, and Hartmut Lutz. 2005. *The Diary of Abraham Ulrikab.* Ottawa: University of Ottawa Press.

Vail, Leroy, and Landeg White. 1991. *Power and the Praise Poem: Southern African Voices in History.* London: James Currey.

Van Cauteren, Willy. 1919. *La guerre et la captivité: Journal d'un prisonnier de guerre en Allemagne.* Brussels: Libraire Nationale d'Art et d'Histoire.

Vansina, Jan. 1990. *Paths in the Rainforest: Toward a History of Political Tradition in Equatorial Africa.* Madison: University of Wisconsin Press.

Venkatachalam, Meera. 2012. "Between the Devil and the Cross: Religion, Slavery, and the Making of the Anlo-Ewe." *Journal of African History* 53: 45–64.

von Hornbostel, Erich Moritz. [1905] 1986. "Die Probleme der vergleichenden Musikwissenschaft" (1905). In *Tonart und Ethos: Aufsätze zur Musikethnologie und Musikpsychologie,* edited by Christian Kaden and Erich Stockmann, 33–48. Leipzig: Reclam.

von Hornbostel, Erich Moritz. 1928. "African Negro Music." *Journal of the International African Institute* 1 (1): 30–62.

von Hornbostel, Erich Moritz, and Otto Abraham. 1904. "Über die Bedeutung des Phonographen für die vergleichende Musikwissenschaft." *Zeitschrift für Ethnologie* 36 (2): 222–36.

von Luschan, Felix. 1896. *Instruktion für ethnographische Beobachtungen und Sammlungen in Deutsch-Ostafrika.* Berlin: Mittler.

von Luschan, Felix. 1904. "Einige türkische Volkslieder aus Nordsyrien und die Bedeutung phonographischer Aufnahmen für die Völkerkunde." *Zeitschrift für Ethnologie* 36 (2): 177–202.

von Luschan, Felix. 1906. "Bericht über eine Reise in Südafrika." *Zeitschrift für Ethnologie* 38 (6): 863–95.

von Luschan, Felix. 1922. *Völker, Rassen, Sprachen.* Berlin: Welt-Verlag.

von Luschan, Felix, and Hermann Struck. 1917. *Kriegsgefangene: Ein Beitrag zur Völkerkunde im Weltkriege; Einführung in die Grundrisse der Anthropologie.* Berlin: Dietrich Reimer.

von Rentzell, Werner. 1922. *Unvergessenes Land . . . Von glutvollen Tagen und silbernen Nächten in Togo*. Hamburg: Alster.

von Tiling, Maria. 1918/19. "Die Vokale des bestimmten Artikels im Somali." *Zeitschrift für Kolonialsprachen* 9: 132–66.

von Tiling, Maria. 1925. *Somali Texte und Untersuchungen zur Somali Lautlehre*. Berlin: Reimer.

Walcott, Derek. 1990. *Selected Poems, 1948-84*. New York: Noonday.

Weidmann, Amanda. 2015. "Voice." In *Keywords in Sound*, edited by David Novak and Matt Sakakeeny, 232–41. Durham, NC: Duke University Press.

Weninger, Josef. 1918. "Anthropologische Untersuchungen indischer und afrikanischer Völkerschaften in deutschen Kriegsgefangenenlagern im Sommer 1917." *Mitteilungen der k.u.k. Geographischen Gesellschaft in Wien* 61 (11): 545–62.

Weninger, Josef. 1927. *Eine morphologisch-anthropologische Studie: Durchgeführt an 100 westafrikanischen Negern, als Beitrag zur Anthropologie von Afrika*. Vienna: Verlag der Anthropologischen Gesellschaft Wien.

Westermann, Diedrich. 1902. "Beiträge zur Kenntnis der Yehwesprache in Togo." *Zeitschrift für Afrikanische un Oceanische Sprachen* 6: 261–90.

Weule, Carl. 1915. "Die farbigen Hilfsvölker unserer Gegner, eine ethnographische Übersicht." Parts 1 and 2. *Kosmos* 12 (6–7): 205–6 and 249–53.

Yamomo, meLê. 2018. *Theatre and Music in Manila and the Asia Pacific, 1869-1946: Sounding Modernities*. Cham, Switzerland: Palgrave Macmillan.

Yamomo, meLê. 2021. "Acoustic Epistemologies and Early Sound Recordings in the Nusantara Region: Phonography, Archive, and the Birth of Ethnomusicology." In *Made in Nusantara: Studies in Popular Music*, edited by A. Johan and M. Santaella, 75–82. London: Routledge, Taylor and Francis.

Yamomo, meLê, and Barbara Titus. 2021. "The Persistent Refrain of the Colonial Archival Logic/Colonial Entanglements and Sonic Transgressions." *World of Music* 10 (1): 39–70.

Zahn, Franz Michael. 1895. "Muttersprache in der Mission." *Allgemeine Missionszeitschrift* 1895: 337–60.

Zedelmaier, Helmut. 2007. "Das Geschäft mit den Fremden: Völkerschauen im Kaiserreich." In *Das "lange" 19. Jahrhundert: Alte Fragen und neue Perspektiven*, edited by Nils Freytag and Dominik Petzhold, 183–200. Munich: Herbert Utz.

Ziegler, Susanne. 2006. *Die Wachszylinder des Berliner Phonogramm-Archivs*. Berlin: Ethnologisches Museum Berlin.

Zimmermann, Andrew. 2001. *Anthropology and Antihumanism in Imperial Germany*. Chicago: University of Chicago Press.

Zimmermann, Andrew. 2015. "Colonization of Antislavery and the Americanization of Empires: The Labor of Autonomy and the Labor of Subordination in Togo and the United States." In *Making Empire Work: Labor and United States Imperialism*, edited by Daniel E. Bender and Jana K. Lipman, 267–88. New York: New York University Press.

Index

Images in plates are indicated by *p1, p2, p3*, etc.

absence, 23, 114, 120, 169n47; of documentation, 26–27; in recordings, 113; of semantic content, 36–37, 53; of World War I in recordings, 41–42, 72
accents, of speakers, 25, 55, 122, 166n28
access: to archives, 151, 154–56; digitization and, 46, 155–56; translations and, 45
acoustic fragments, 15–18, 89, 102–3, 117; speakers and, 19–21
Afelevo, Hiob Kwadzo, 124–25, 178n42
African soldiers, 4–6, 27–28, 49, 133; violence and, 113
agency, 129; genres and, 44; narrative, 16. *See also* authorship
anthropology, 106–7, 174n8; race and, 31–33
anthropometry: colonialism and, 84–85; Doegen and, 22, 58–59; Farnana and, 61, 169n48; Niang, A., and, 60–62; POWs and, 4, 168n40; race and, 35, 62–63, 174n8; recordings and, 32–33; speakers and, 165n18; violence and, 152–53; in Wünsdorf, 59–60
Apoh, Wazi, 135, 141, 143–44
archival traces, 3–6, 17–20; of Bischoff, 131–35; of Niang, A., 24–25, 48; of Nur, 76–77; of speakers, 153–54
archives, 11–12, 33, 148–49; access to, 151, 154–56; colonial knowledge production and, 87–88; languages in, 167n34; networks of, 3–4, 48–49, 59–60; oral, 16, 18, 82, 87, 89; politics of, 24–25. *See also* colonial archive
army: Belgian, 12, 61–62, 101, 104, 174n1; French, 25–27, 49, 63, 150
asemantic voice, 53–54

authorship, 76, 86, 89, 154; colonial knowledge production and, 114–15

barriers, language, 37, 39, 49, 56, 129–30
Baule/Baoulé (language), 61, 167n36
Belgian army, 12, 61–62, 101, 104, 174n1
Belgian Congo, 6, 108–9, 117–18; Kilo, 103–4, 107, 116, *p12*
Belgium, 104, 107, 117–18, 174n5
ben Ahmad, Asmani (speaker), 42, 64–65, 174n3, *p3*
Berbera, Somalia, 81–82, 91–94, 173n39
Berlin, Germany: Luna Park, 33, 81–83, 86, *p7, p9*; Odeon record company in, 12, 25, 51; Phonogramm-Archiv, 31–33, 82–83, 161n5, 162n10, 164n13, 167n35; World War II and, 170n6. *See also* Lautarchiv, Berlin
bin Bedja, Ali (speaker), 42, 66, 94
Bira (language), 102–3, 108–9, 111, 115–16; genres of, 106–7
Birdsall, Carolyn, 149
Bischoff, Stephan (speaker), 6, 12, 101; archival traces of, 131–35; evangelism and, 118, 138–40; fable of, 135, 138–40, 145; as language assistant, 131–32; languages of, 125–27; meaning and, 17; Meinhof and, 118–20, 123, 125–27, 131–38; migration of, 131, 141; personal files of, 119, 121, 131–33; recordings of, 120, 122–24, 178n41; spirit language and, 128–29; voice of, 120
Bleek, Wilhelm, 34–35, 165n23, 177n37
bodies, 30, 41; documentation and, 62; exoticism and, 78; knowledge production and, 60; life casts of, 33, 48, 60–62, 80, 152, 169n48, 171n17; race and, 12–13, 70–71, 78; Wünsdorf and, 67

Bose, Fritz, 53–54, 151, 168n39
Britton, Cecilia, 102
Brosens, Griet, 104, 174n1, 174n5
Burkina Faso. *See* Dahomey
Buru (speaker), 49–50, 61, 167n36

Central Powers, 1, 58, 168n44
Césaire, Aimé, 148
"Chandra's Death" (Guha), 11–12, 45
Christianity, 124; languages and, 118, 120, 122–23
close listening, 43, 47–48, 152–53
collaboration, 31, 71, 74–76, 81, 86; listening and, 151, 155
collections, 3, 77, 176n14; colonialism and, 6–7, 46, 161n5; digitization and, 164n13; exoticism and, 97, 113; of languages, 38, 43–44, 113; of recordings, 5, 31
colonial archive, 14, 23, 37, 46, 77, 150, 155; exoticism and, 81–82; Lautarchiv, Berlin, and, 12–13; power and, 71–72; ritual and, 17; violence and, 113–14
colonial films, 111, 176n19
colonial history, 5, 13; speakers and, 30, 144–45
colonialism: anthropometry and, 84–85; collections and, 6–7, 46, 161n5; exploitation and, 32; linguists and, 34, 124, 144, 148; power and, 19, 143–45; spirituality and, 140–44; violence and, 107–8, 117, 180n62; voice and, 147
coloniality, 13, 25, 34, 45–46, 77, 79, 86, 129, 145
colonial knowledge production, 8, 24–25, 128–29, 147, 149; archives and, 87–88; authorship and, 114–15; communication and, 43, 87, 152; debris of, 4, 12–13, 18, 45, 60, 114; KPPK and, 33–34; networks of, 174n3; recordings and, 45
colonial politics, languages and, 163n4
colonial soldiers, 6, 26–28, 59, 79
communication, 125, 140; colonial knowledge production and, 43, 87, 152; drum languages and, 102, 108, 110–12, 176nn16–17; failure of, 37, 48, 56–57, 110, 122; linguists and, 120, 122; listening

and, 152; missionaries and, 102, 122–23; recordings and, 38–39; speakers and, 50, 54–55; urgency and, 51–52
Comoros, 31, 64–65, 167n32, 167n34
Congo, Belgian, 6, 108–9, 117–18
connections, in documentation, 58–59
conscription, 1, 27–28, 61–62, 73
context, 30, 69, 72, 164n7; fragments and, 17, 46, 117, 148; translation and, 107, 149
critical listening, 147, 149–50, 155

Dahomey, 1, 49, 81, 128, 146
Das, Santanu, 26–27
death, 81; from tuberculosis, 63, 169n46; violence and, 27
debris, of colonial knowledge production, 4, 12–13, 18, 45, 60, 114
decolonization, 39, 46, 145, 154–56
deities, in recordings, 17, 124–25, 127, 140; drum languages and, 128–29
deportation, 14, 62, 104; to Romania, 48, 52, 54–55, 57–60, 63, 169n46
Der Krieg und die Grammatik: Ton- und Bildspuren aus dem Kolonialarchiv (War and Grammar: Audiovisual traces from the colonial archive) (exhibition), 7, 22, 82, 149–50, *p16*
Dervish movement, 92–93, 95, 98, 147–48
Diallo, Mamadou Samba (speaker), 9, 41, 53
Diallo, Samba (speaker), 7–8, 26–27, 61; Lautarchiv and, 9, 28
difference, racial, 4, 35, 70–71, 80, 84
digitization, 13–14, 173n34; access and, 46, 155–56; collections and, 164n13
documentation, 15; absence of, 26–27; of Bischoff, 119, 121; connections in, 58–59; of Kudjabo, 103–4; linguists and, 43; of Niang, A., 41, 62–63; recordings and, 28–30, 164n11; registration and, 40–41, 167n31; translation of, 156
Doegen, Wilhelm, 18, 30–32, 151–52, *p5*; anthropometry and, 22, 58–59; communication and, 39; drum languages and, 111–15; on Niang, A., 25–26, 53; Nur and, 87, 89–91; and POW camps as "field

202 INDEX

site," 35; race and, 54; radio and, 112–15; registration and, 41–42
Dolar, Mladen, 122
double inscription, 24, 45–46, 87–88, 115
drum languages, 174n3, 177n21; in colonial films, 111, 176n19; communication and, 102, 108, 110–12, 176nn16–17; deities and, 128–29; Doegen and, 111–15; Kudjabo and, 109–15
drums, 12, 103, 174n2, 174n8, 176n14, *p11*
Du Bois, W. E. B., 177n32

earwitnessing, 148–49, 156
Eschenberg, Myron, 27–28
echoes, 5–6, 15–19, 28, 101, 147–48, 156
Egeh Gorseh, Hirsi, 81–82, 87, 150
Eickstedt, Egon von, 53, 84–85, 168n40
enemies, 34, 39, 67, 70
Engländerlager (English camp). *See* Ruhleben (internment camp)
English (language), 69
Escaped! (Ellison), 67, 69, 71
ethnic shows. *See* Völkerschauen
Ettinghausen, Maurice L., 67–68, 170n5
evangelism: Bischoff and, 118, 138–40; Meinhof and, 110, 118, 120, 126; translations and, 122–23
Ewe (language), 73, 120, 122–27, 130–32, 138
exhibitions, 77, 149, *p16*; recordings and, 6–8, 22, 37–38, 167n7; translation and, 150–51
exoticism, 7; bodies and, 78; collections and, 97, 113; colonial archive and, 81–82; otherness and, 67, 83–84, 97; POWs and, 79; race and, 95–96; translation and, 151
expertise, 48, 74, 102, 113–15, 134, 148
exploitation, 77–78; colonialism and, 32; Meinhof and, 34; Nur and, 86; of POWs, 24, 30; *Verwertungsmaschine* and, 58–59, 71, 79, 86
extractive knowledge production, 153–54

fables, 118, 120, 141, 144, 180n62; of Bischoff, 135, 138–40, 145

failure, of communication, 37, 48, 56–57, 110, 122
Farnana, Paul Panda, 62, 117–18, 162n9, 177n32; anthropometry and, 61, 169n48
Finnegan, Ruth, 108
formats, recording sessions and, 45, 167n35
Foucault, Michel, 16, 19
fragments: acoustic, 15–18, 89, 102–3, 117; context and, 17, 46, 117, 148; narratives and, 11–12, 15, 18–19
"free speech," 13, 36, 50–51
French army, 25–27, 49, 63, 133, 150
Frobenius, Leo, 34, 63, 85, 169n46

gender relations, 116–17
genres, 93, 147, 173n37; agency and, 44; of Bira, 106–7; fable and, 135; Meinhof and, 50; musical, 2–3; poetry and, 88–89
German Togoland, 131–33, 177n34, 179n57; Krachi, 118, 135, 140–45, 179n59, *p15*
Germany, 3–4, 15, 59–60, 67, 80–82, 130; Hamburgisches Kolonialinstitut, 6, 73–74, 123, 131–32; radio broadcasts in, 103, 111–15, 177n22. *See also* Berlin, Germany
Glissant, Édouard, 102, 128
Goebbels, Heiner, 152–53
grammars, 4, 77; linguists and, 42, 71–72
Gregoire, Mamadou (speaker), 145–46
Grohs, Albert, 70, 170n5
Gruner, Hans, 143–44, 179n59
Guha, Ranajit, 15, 71, 76; "Chandra's Death," 11–12, 45

Habil, Ali Jama, 91, 95
Hagenbeck, Carl, 33, 80–82, 85, 97, 172n23
Halbmondlager (Half Moon camp). *See* Wünsdorf (POW camp)
Hambruch, Paul, 31, 65, 110–11, *p3*
Hamburgisches Kolonialinstitut, 6, 73–74, 123, 131–32
Hamilton, Carolyn, 149
Haraway, Donna, 3, 8
Hassan, Mohamed 'Abdille, 91–92, 94, 98
Heepe, Martin, 167n34

INDEX 203

Heinitz, Wilhelm, 109, 112
hierarchies, of race, 96, 173n43
historiography, 3, 76; of World War I, 23, 26–27
Holz, E., 82–83
Honour, Hugh, 96–98
human remains, 33–34, 145, 171n17, 172n23; Lautarchiv, Berlin, and, 165n23
Humboldt Box, 3, 23, 37, 163n1
Humboldt Forum, 3, 13, 149, 161n5, 162nn6–7

images: of Nur, 77–79, 95–98, 173n42; race and, 153; of speakers, 58, 62
informants, 6, 28, 71, 75–76, 101–2, 132–34, 170n7
inscription, double, 24, 45–46, 87–88, 115
internment camp: Ruhleben, 36, 100, 131–32, *p4–5*
Irvine, Judith, 55–56
isiXhosa (language), 5, 89; Ntwanumbi and, 42, 70, 98–100, 120, 140, 151, 166n28

Jámafáda (speaker), 1–3, 21, 150; narrative of, 22, 42–43; personal file of, 20, 22, 49; war and, 58

Kaiserreich, 59, 79, 86
Kane, Fatou Cissé (translator), 5, 150–51, 155–56, 169n51
Kilo, Belgian Congo, 103–4, 107, 116, *p12*
knowledge production: bodies and, 60; extractive, 153–54; informants and, 75–76; personal files and, 52–53; in POW camps, 30–35, 42, 78–79; *Völkerschauen* and, 80. *See also* colonial knowledge production
Kollwitz, Käthe, 85
Kolonialinstitut, Hamburgisches, 6, 73–74, 123, 131–32
KPPK (Königlich-Preußische Phonographische Kommission) (Royal Prussian Phonographic Commission), 5, 15, 24, 113; colonial knowledge production and, 33–34; Kaiser Wilhelm II and, 31; Lus-

chan and, 31–33; nationalism and, 84–85; recordists of, 14, 44, 48–49, 56; resistance to, 66–67, 69; secrecy and, 35–36, 102; speech acts and, 166n28
Krachi, German Togoland, 118, 135, 140–45, 179n59, *p15*
Kudjabo, Albert (speaker), 6, 12, 101–2, 150, *p11*, *p13–14*; documentation of, 103–4; drum languages and, 109–15; languages of, 103; migration of, 104, 107–8, 174n1; personal file of, 105; radio and, 112–15; recordings of, 107–8, 111, 115–17, 175n11; songs of, 104, 106, 108; stories of, 115–16

Landau, Paul, 123
Lange, Britta, 3, 5, 32, 147
language assistants, 123, 178n39; Bischoff as, 131–32; Nur as, 73–74, 78–79, 96, 167n32
language barriers, 37, 39, 49, 56, 129–30
languages, 35; in archives, 167n34; Baule, 61, 167n36; of ben Ahmad, 64–65; Bira, 102–3, 108–9, 111, 115–16; of Bischoff, 125–27; Christianity and, 118, 120, 122–23; collection of, 38, 43–44, 113; colonial politics and, 163n4; English, 69; Ewe, 73, 120, 122–27, 130–32, 138; isiXhosa, 5, 89; of Kudjabo, 103; migration of, 117, 126, 130; Mòoré, 21–22, 49–50, 58, 151, 162n10; "mother tongue" and, 125–26, 178n43; music and, 12–13; opacity of, 71, 102, 110–12, 125, 128–29; race and, 165n23; registration and, 42–43; resistance and, 125–26, 176n46; secret, 117, 124–27, 129–30, 142; Somali, 4, 72, 74–78, 86–89, 94; spirit, 101–2, 124–26, 142; surveys of, 38, 167n30; tonal, 39, 102, 108, 111, 114, 122; Wolof, 1, 23, 25–28, 51–52, 55–57, 169n51; Yevegbe, 101–2, 124–28, 178n48. *See also* drum languages
langue, parole and, 37, 166n27
larynges, 35, 78, 165n23, 171n17
Lautarchiv, Berlin, 3–5, 24, 162n6; colonial archive and, 12–13; Diallo and, 9, 28; digitization in, 173n34; documentation

in, 40–41; human remains and, 165n23; at the Humboldt Box 3, 23, 37, 153, 163n1; at the Humboldt Forum, 149, 162n7; as oratory, 6–7, 28, 47; recordings and, 19–21

Le Guin, Ursula K., 3

life casts, of bodies, 33, 48, 60–62, 80, 152, 169n48, 171n17

linguistic records, 18–19, 161n4; semantic content and, 14–15, 76, 163n3

linguistics, 32; missionaries and, 124, 126–27, 138–40; publications for, 42; race and, 19, 111–12; war and, 72

linguistic specimens, 3, 25, 87–88, 94

linguists, 14, 28, 161n4; colonialism and, 34, 124, 144, 148; communication and, 120, 122; documentation and, 43; grammars and, 42, 71–72; narratives and, 117; opportunism of, 3, 8, 12, 32–33, 72, 154; semantic content and, 15

listening, 47; collaboration and, 151, 155; communication and, 152; critical, 147, 149–50, 155; recordings and, 48; speech acts and, 37, 44

Lloyd, Lucy, 177n37

loss, 11, 43, 64–65, 107

Luna Park, Berlin, 33, 81–83, 86, *p7*, *p9*

Luschan, Felix von, 163n3; documentation of, 164n11; KPPK and, 31–33; Meinhof and, 56–58, 109–10, 168n45; Pöch and, 59

Maharero, Friedrich, 81, 172n23

Marchal, Jules, 107

Mbembe, Achille, 70–71

meaning, 43, 108, 140, 163n3; Bischoff and, 17; Niang, A., and, 42; sound and, 13–14, 37, 122; translations and, 50; voice and, 16, 39, 54

Meinhof, Carl, 129–30, *p5*; Bischoff and, 118–20, 123, 125–27, 131–38; drum languages and, 109–11; evangelism and, 110, 118, 120, 126; Ewe and, 122–23; exploitation and, 34; genres and, 50; Jámafáda and, 22; KPPK and, 31–32; Luschan and, 56–58, 109–10, 168n45; Niang, A., and, 25; Nur and, 73–74

Mendi, SS, 2, 18, 161n3

men of color, 70, 133; as specimens, 35

Meyer, Birgit, 126–27

migration, 3–4, 6, 95, 148, 156; of Bischoff, 131, 141; of Kudjabo, 104, 107–8, 174n1; of language, 117, 126, 130; *Völkerschauen* and, 82–84

missionaries, 12, 144–45; communication and, 102, 122–23; linguistics and, 124, 126–27, 138–40; schools of, 101, 103–4, 127, 131–32

Mòoré (language), 21–22, 49–50, 58, 151, 162n10

Morrison, Toni, 11, 66

"mother tongue," 125–26, 178n43

Münster (POW camp), 101, 103, 108–9, 167n32, 175n11, *p11*

music, 153; genres of, 2–3; languages and, 12–13; race and, 109

musicology, 25, 53; drum language and, 176n17

names, of POWs, 161n4, 162n1, 169n2

narratives, 6–7, 14, 48–51, 106–7, 141–43; agency and, 16; of ben Ahmad, 64–65; fragments and, 11–12, 15, 18–19; of Jámafáda, 22, 42–43; linguists and, 117; of Nur, 76, 96; recordings and, 24–25

nationalism, 27, 91, 130; KPPK and, 84–85

National Socialist (NS) state, 168n39–40. *See also* Nazism

Nazism (National Socialism), 149, 168n39–40, 173n43

networks: of archives, 3–4, 48–49, 59–60; of colonial knowledge production, 174n3

Niang, Abdoulaye (speaker), 1, 23, 161n4, 163n5, *p1–2*; anthropometry and, 60–62; archival traces of, 24–25, 48; documentation and, 41, 62–63; Doegen and, 25–26, 53; personal file of, 29, 51; pleas of, 51–60; recording of, 25–28, 47–48, 63–64; songs of, 25–27, 42, 49–51; translations of, 26–27, 169n51; voice of, 39–40, 53

Niang, Serigne Matar (translator), 25–26, 47, 51, 56, 163n5

INDEX 205

NS (National Socialist) state, 168n39–40. *See also* Nazism

Ntwanumbi, Josef (speaker), 6, 12, 161n4; isiXhosa and, 42, 70, 98–100, 120, 140, 151, 166n28; recordings of, 120, 164n13; songs of, 42, 50, 108

Nur, Mohamed (speaker), 6–7, 12, 66, 150, 161n4, *p4–6, p9*; archival traces of, 76–77; Doegen and, 87, 89–91; Egeh Gorseh and, 82; images of, 77–79, 95–98, 173n42; as language assistant, 73–74, 78–79, 96, 167n32; narratives of, 76, 96; poetry of, 92–95; recordings of, 87–89, 173n35, 173nn39–41; Ruhleben and, 75–79, 87–95, 132; semantic content and, 36; Slevogt and, 95–98, 171n16; songs of, 90–91; *Völkerschauen* and, 86–87; von Tiling and 72–75, 166n27

objectification, 59, 79–81, 147
Odeon record company, Berlin, 12, 25, 51
opacity, of language, 71, 102, 110–12, 125, 128–29
opportunism, of linguists, 3, 8, 12, 32–33, 72, 154
oral archives, 16, 18, 82, 87, 89
oratory, Lautarchiv as, 6–7, 28, 47
orature, 17–18; politics and, 91; power and, 135, 138; Somali, 36
Ossey, Johannes (translator), 118, 124, 135, 138–39, 150–51, 178n41
otherness, 106–7, 111, 128; exoticism and, 67, 83–84, 97
Ottoman Empire, Triple Entente and, 168n44

Panconcelli-Calzia, Giulio, 78, 109
parole, langue and, 37, 166n27
Peeren, Esther, 164n7
performativity, 17–18, 39, 43–46, 53–54, 56, 140
performers, of *Völkerschauen*, 32–33, 79–84, 87, 95–96
personal files, 28, 30, 38–42, 44; of Bischoff, 119, 121, 131–33; of Jámafáda, 20, 22, 49;

knowledge production and, 52–53; of Kudjabo, 105; of Niang, A., 29, 51

Phonogramm-Archiv, Berlin, 31–33, 82–83, 161n5, 162n10, 164n13, 167n35
Phonogrammarchiv, Vienna, 28, 30, 32
photographs, 143; of POWs, 177n25, 177n36; secrecy and, 67–68. *See also* images
Pöch, Rudolf, 33–35, 61–62, 165n18–19, 165n23; Luschan and, 59
poetry: genres and, 88–89; of Nur, 92–95; war in, 145–46
politics, 93–94, 98, 163n4; of archives, 24–25; orature and, 91
portraits, of speakers, 7, 37–38, 78, 85, 145, 166n29
POW camps: knowledge production in, 30–35, 78–79; Münster, 101, 103, 108–9, 167n32, 175n11, *p11*; recording sessions and, 40, 76–77, 167n31; Soltau, 12, 101, 109; Turnu Măgurele, 61–63, *p2*; *Völkerschauen* and, 86–87; Wünsdorf, 22, 47–48, 58, 61, 145–46, *p3*. *See also* Ruhleben (internment camp)
power, 81, 125–26; acoustic fragments and, 18–19; colonial archive and, 71–72; colonialism and, 19, 143–45; orature and, 135, 138
Prakash, Gyan, 44, 144
prisoners of war (POWs), 6, 8, 28, 70, 103, *p3*; anthropometry and, 4, 168n40; exoticism and, 79; exploitation of, 24, 30; names of, 161n4, 162n1, 169n2; photographs of, 177n25, 177n36. *See also* speakers
propaganda, 4–6, 59, 70–71, 78–79, 133; publications and, 34–35, 85
publications, 113–14; linguistic, 42; propaganda and, 34–35, 85; war and, 72–73

race, 1, 4; anthropology and, 31–33; anthropometry and, 35, 62–63, 174n8; bodies and, 12–13, 70–71, 78; exoticism and, 95–96; hierarchies of, 96, 173n43; images and, 153; languages and, 165n23; linguistics and, 19, 111–12; music and, 109; voice and, 53–54, 168n41

racial difference (alleged), 4, 35, 70–71, 80, 84
racism, 1, 54, 69–70, 133, 151, 173n43, 178n38
radio broadcasts, German, 103, 111–15, 177n22
recordings, 1–3, 12, 16, 18–21, 24, 167nn34–35; absence in, 113; anthropometry and, 32–33; of Bischoff, 120, 122–24, 178n41; collections of, 5, 31; colonial knowledge production and, 45; communication and, 38–39; deities in, 17, 124–25, 127, 140; documentation and, 28–30, 164n11; exhibitions and, 6–8, 22, 37–38, 167n7; "free speech" in, 13, 36, 50–51; of Kudjabo, 107–8, 111, 115–17, 175n11; listening and, 48; of Niang, A., 25–28, 47–48, 63–64; of Ntwanumbi, 120, 164n13; of Nur, 87–89, 173n35, 173nn39–41; in Ruhleben, 66, 87–95, 169n1; secrecy of, 69–70; *Völkerschauen* and, 32–33, 82–84; von Tiling and, 170n10; war in, 36, 41–43, 73, 108, 120, 122
recording sessions, 28, 36, 44, 49–51, 83; formats and, 45, 167n35; POW camps and, 40, 76–77, 167n31
recordists, of KPPK, 14, 44, 48–49, 56
record keeping, 15, 36–37; genres of, 89; linguistic, 161n4
recruitment, 1, 22, 36–37, 58, 146; from Senegal, 25–28, 164n10; songs and, 53
registers, of speech, 39–40, 53–56, 166n28
registration: documentation and, 40–41, 167n31; language and, 42–43
Rentzell, Werner von, 141–43, 179n57
repatriation, 145, 161n5
repertoires, 18–19, 23–24, 45–46, 84, 152
representation, 1, 4–5, 27, 60–61, 76, 97
resistance: to KPPK, 66–67, 69; languages and, 125–26, 178n46; to *Völkerschauen*, 81
ritual, colonial archive and, 17
Roberts, Toby, 67–71
Romania: deportation to, 48, 52, 54–55, 57–60, 63, 169n46; Turnu Măgurele POW camp in, 61–63, p2
Rooble, Shire (speaker), 83–84, 98

Royal Prussian Phonographic Commission. *See* KPPK (Königlich-Preußische Phonographische Kommission)
Ruhleben (internment camp), 36, 100, 131, p4–5; Ellison and, 66–70; Nur and, 75–79, 87–95, 132; recordings in, 66, 87–95, 169n1

Samatar, Said Sheikh, 88, 91, 95, 172n27, 173n39
Samba Diallo, Mamadou. *See* Diallo, Mamadou Samba
Schebesta, Paul, 174n8
schools, missionary, 101, 103–4, 127, 131–32
Schünemann, Georg, 9, 31, 146, 164n13, 173n33
secrecy: KPPK and, 35–36, 102; photographs and, 67–68; of recordings, 69–70
secret languages, 117, 124–27, 129–30, 142; translations of, 101–2, 106
semantic content, 148; absence of, 36–37, 53; linguistic records and, 14–15, 76, 163n3. *See also* meaning
Senegal: recruitment from, 25–28, 164n10; speakers from, 55
slave trade, 126, 178n44
Slevogt, Max, p10; Nur and, 95–98, 171n16
soldiers: African, 4–6, 27–28, 49, 133; colonial, 6, 26–28, 59, 79
Soltau (POW camp), 12, 101, 109
Somali (language), 4, 72, 74–78, 86–89, 94; orature of, 36
Somalia: Berbera, 81–82, 91–94, 173n39; Dervish movement in, 92–93, 95, 98, 147–48
Somali "Villages" (Somali Dörfer): at Luna Park, 33, 86–87, p7, p9; in *Völkerschauen*, 80–81, p8
Somali Texte (von Tiling), 72–73, 75–77, 86, 93–94, 98
songs, 2, 12, 83, 172n27, 178n41; of Kudjabo, 104, 106, 108; of Niang, A., 25–27, 42, 49–51; of Ntwanumbi, 42, 50, 108; of Nur, 90–91; recruitment and, 53
sound: meaning and, 13–14, 37, 122; voice and, 39, 53

INDEX

speakers, 51, 156; accents of, 25, 55, 122, 166n28; acoustic fragments and, 19–21; anthropometry and, 165n18; archival traces of, 153–54; ben Ahmad, 42, 64–65, 174n3, *p3*; bin Bedja, 42, 66, 94; Bischoff, 6, 12, 101; Buru, 49–50, 61, 167n36; colonial history and, 30, 144–45; communication and, 50, 54–55; Diallo, 7–8, 26–27, 61; documentation of, 15; as enemies, 39; Gregoire, 145–46; images of, 58, 62; as informants, 6, 28, 71, 75–76, 101–2, 132–34, 170n7; Jámafáda, 1–3, 21, 150; Kudjabo, 6, 12, 101–2, 150, *p11*, *p13–14*; Niang, A., 1, 23, 161n4, 163n5, *p1–2*; Ntwanumbi, 6, 12, 161n4; Nur, 6–7, 12, 66, 150, 161n4, *p4–6*, *p9*; portraits of, 7, 37–38, 78, 85, 145, 166n29; Rooble, 83–84, 98; Sambadialo, 9, 41, 53; voice and, 129

"speaking drum," 12, *p11*

specimens: linguistic, 3, 25, 87–88, 94; men of color as, 35

speech, registers of, 39–40, 53–56, 166n28

speech acts, 12–13, 15–16, 46, 55; KPPK and, 166n28; listening and, 37, 44

Spieth, Jakob, 124–25, 127

spirit languages, 101–2, 124–26, 142; Bischoff and, 128–29

spirituality, colonialism and, 140–44

Spivak, Gayatri Charavorty, 44, 114, 151

Steedman, Carolyn, 106

Sterne, Jonathan, 16, 25

Stoever, Jennifer Lynn, 54

Stoler, Laura, 114, 177n29

Stopa, Roman, 35, 165n23

Storch, Anne, 124, 127–28, 178n38

stories, 3, 16, 38, 42–43, 51, 152–53; of ben Ahmad, 64; of Kudjabo, 115–16

Struck, Herman, 7, 145–46

Stumpf, Carl, 31, 35–36, 113, 166n25, 174n3, 177n26

surveys, of languages, 38, 167n30

Thilenius, Georg, 109

tirailleurs, 1, 14, 23, 41, 145–46

Togoland, German, 131–33, 177n34, 179n57

tonal languages, 39, 102, 108, 111, 114, 122

Toso, Victor, 123, 178n39

traces, archival, 3–6, 17–20

transcription, 38, 109, 121, 144, 166n29, 177n37

translations, 5, 17, 39, 163n3; access and, 45; context and, 107, 149; of documentation, 156; evangelism and, 122–23; exhibitions and, 150–51; listening and, 48; meaning and, 50; of Niang, A., 26–27, 169n51; of Nur, 75; of secret languages, 101–2, 106; war and, 42–44

translators, 115–16; Kane, 5, 150–51, 155–56, 169n51; Niang, S. M., 25–26, 47, 51, 56, 163n5; Ossey, 118, 124, 135, 138–39, 150–51, 178n41; Warsame, 74–75, 83–85, 90–91, 97, 173n35

Triple Entente, 1, 34, 165n20; Ottoman Empire and, 168n44

tuberculosis, death from, 63, 169n46

Turnu Măgurele (POW camp), 61–64, *p2*

Ulrikab, Abraham, 81, 172n23

urgency, communication and, 51–52, 57–59

Venkatachalam, Meera, 126–27, 143–44

versioning, 18, 89, 164n7; historiography and, 26

Verwertungsmaschine, 58–59, 71, 79, 86

Vienna, Phonogrammarchiv, 28, 30, 32

violence, 60, 81, 106; African soldiers and, 113; anthropometry and, 152–53; colonial archive and, 113–14; colonialism and, 107–8, 117, 180n62; death and, 27

voice, 17, 77; asemantic, 53–54; of Bischoff, 120; colonialism and, 147; meaning and, 16, 39, 54; of Niang, A., 39–40, 53; race and, 53–54, 168n41; sound and, 39, 53; speakers and, 129

Völkerschauen (ethnic shows), 85, 114; knowledge production and, 80; migration and, 82–84; Nur and, 86–87; performers of, 32–33, 79–84, 87, 95–96; POW camps and, 86–87; recordings and, 32–33, 82–84; Somali Dörfer and, 80–81, *p7–9* von Tiling, Maria: Nur and, 72–75, 166n27; recordings and, 170n10; *Somali Texte*, 72–73, 75–77, 86, 93–94, 98

war, 2, 4; Jámafáda and, 58; linguistics and, 72; in poetry, 145–46; publications and, 72–73; in recordings, 36, 41–43, 73, 108, 120, 122; translations and, 42–44
Warsame, Bodhari (translator), 74–75, 83–85, 90–91, 97, 173n35
Weninger, Josef, 9, 49, 59–63, 167n36, 168n41
Westermann, Diedrich, 142, 178n48
Wilhelm II, Kaiser, 8, 35, 150; KPPK and, 31

Wolof (language), 1, 23, 25–28, 51–52, 55–57, 169n51
World War I, 1, 3–5, 156; absence of, 41–42, 72; historiography of, 23, 26–27; Nur and, 75; Triple Entente and, 165n20
World War II, 112, 170n6
Wünsdorf (POW camp), 22, 47–48, 58, 61, 145–46, *p3*; anthropometry in, 59–60; bodies and, 67; linguists and, 14

Yamomo, meLê, 154–55
Yevegbe (language), 101–2, 124–28, 178n48

Zeller, Rudolph Jacob, 77–78, 171n15, *p6*